Revisiting Pakistan's National Security Dilemma

Edited by:

Prof. Dr. Iram Khalid

PEACE PUBLICATION
Manzoor Manzil 42 Urdu Bazar, Lahore

This edition, jointly published by:

Peace Publication
42. Manzoor Manzial Opp Govt Muslim
Model High School Urdu Bazar Lahore
Ph: 0322-4298311-03074505020
E-mail: peace_ publication @yahoo.com

Al-Abbas International
Publisher & Book Sellers, Lahore
E-mail:alabbasbooks@gmail.com
Web: www.alabbasbooks.pk
Mob: 0333-4172010

Distributor:

Kitab Waadi
42. Manzoor Manzial Opp Govt Muslim
Model High School Urdu Bazar Lahore
Ph: 042-37361448, Mob: 0322-4011864
E-mail: kitabwaadi@yahoo.com

Royal Book Company
Bg-5, Rex Centre Fatima Jinnah
Road Saddar Karachi.
E-mail: royalbook@hotmail.com
Web: www.royalbook.com.pk

First Published 2016

Printed by: Haji Hanif Lahore, Pakistan

Binder: Qayyum Ahmad

ISBN: 978-969-9988-27-1

PREFACE

In the current age of globalization and exceedingly progressive economics and democracies, Pakistan has still a flawed structure of democracy and is still hung on the lack of national security, even after 68 years of its inception. Today, Pakistan faces multifaceted and multilayered challenges which are multifarious in magnitude and complexity, that makes them highly volatile and mark them high on the scale of un-resolvability. Such challenges are internal as well as external in nature. The internal challenges include flawed democracy, estranged society, internal disharmony, fractured state, decorticated institution, corrupted bureaucracy, bad governance, sectarianism, crippling economy, dynastic politics, frail political leadership, radicalism, military dominance and weak judicial system. The corruption on all levels and in every political and governmental establishment is deep-rooted and there is a little to none hope for its eradication in any near future. Including all high-level institution, the system is tarnished from within and bad governance has become a customary norm in political realm of Pakistan.

Pakistan faces most daunting issues from within. It is also trying to cope with existential crisis as the writ of government is challenged from variety of radical school of thoughts, ranging from rightist Islamic extremist to the leftist communal and ethnic identities proclamation to the separatist movements. In the meantime, the global war on terrorism and Pakistan's direct and active involvement in it lingers to cripple already daunting state of Pakistan.

The principle challenges for Pakistan to overcome (on internal as well as external front), as early as possible, in order to sustain, secure and develop in ever changing and developing 21st century are; stimulating the failing economy, assembling defiance mechanism to combat militancy and extremism, resolving the energy crisis, introducing the educational reform, providing basic necessities of life, improving daily wages average for workers, check and balance on radical Islamist groups and generating jobs to engage the highly talented yet unemployed bulge of youth of population to ward off an overwhelmingly apparent and imminent catastrophe. The implications of all these problems for national security are obvious and can be ignored, if Pakistan wants to survive.

Increasing the hard luck of Pakistan, the internal societal and institutional shortcoming are not the only challenging problems. Rather the geographical location, highly unpredictable neighborhood and the main malfunction of geopolitics and discrepancies of regional and international actor, not only make Pakistan strategically significant but also exposes it to limitless vulnerabilities and security dilemmas. The burden of history have consistently put security at the top of Pakistan's national agenda, and it hasn't changed yet. At present Pakistan confronts a number of conventional and asymmetric, traditional and non-traditional challenges, hard and soft threats.

Now, on the front of external threats to Pakistan national security, USA, as in international actor and India, as in regional actor are looming threats, and in that very order.

America's pivot is also momentous to Pakistan's national security challenges. This power altering axis is extensively observed here in Pakistan, as elsewhere in Asia, which is intended to deal with terrorist safe haven in Afghanistan and Pakistan, containment of China's rise as a power and economics giant, via back door politics as, on the international public front America is pursuing a hand forward for economic commitment and partnerships with Beijing, simultaneously.

In the same regard externally, Pakistan's identical regional challenges encompass its associations with boarding countries of India and Afghanistan. Both fronts have been a source of ever constant over looming security dilemmas, for past couple of decades now. That's one of the reasons that Pakistan has strained to being get involved in straight antagonizing situation on these two regional fronts. Bearing in mind Pakistan's strategic and geographical value, global environment needs to be considered. This is disconcerted and in flux, with global power hubs and policy shifts in direct relationship with the developing countries with their interests at stake, throughout the world. The predominant trend is inclination to compete rather than cooperate. Tension among major powers has, in fact, intensified the threat to global instability including national securities of periphery states.

The all three chief global politics altering changes of 21st century have direct insinuations on Pakistan's national security. One, the increasing uproar and upheaval in the oil rich Middle East; two, the bright prospects of China's rise to counter US in the race of hegemony, and the US increasing hinge to Asia, which has opened new gates of opportunities for Asian states and has also gave birth to new dangerous Great Game on this continent.

The next insecurity for Pakistan is nuclear dilemma. Besides being a center of allegations from international community for illegitimately making nuclear deals with Middle Eastern states, the nuclear arsenal of

Pakistan are in threat from outside, terrorist and extremist groups, further increasing Pakistan's struggles.

Pakistan faces a full gamut of security challenges, with highlights on terrorism and militancy. Conventional threats from powerful neighbors and the imperative to retain and sustain its nuclear deterrence integrity rank high on the spectrum of imminent security challenges. Indeed, the amalgamation of terrorist and trivial regional challenges has induced Pakistan to function efficiently on two grounds, simultaneously; Cope with internal and external challenges. The functionalism of this dual engagement continues to pretense tough choices for Pakistan in response to a changing threat environment.

ACKNOWLEDGEMENT

Alhamadolillah, with the grace of Allah Jalahshanuhu the research work has been completed. I am extremely thankful to Allah for all that has contributed to the completion of this work. Without his will this work would surely not have become a reality. For me these are the moments of thankfulness to Allah rather than personal fulfillment. With feelings of gratitude to Allah I acknowledge that nothing in this world belongs to me, everything belongs to Allah and is destined to return to Him even my own self.

I owe a very special department of gratitude to my honorable colloquies and dear students and for their contribution, Dr. Rehana Saeed Hashmi, Mr. Ali Naqvi, Col. Junaid, Maryam Kamal, Zirwa Ahmed, Zainab Ahmed, Arooj Naveed, Salman Ahmed and Sana Shoukat

I am really thankful to Sohaib Rehman for his contribution in proof reading and help in editing the draft. I owe my thanks to Zakia Bano and Farah for their compilation of details about the operation Zerb-e-Azab. I am also grateful to Komal, Mehwish, Zonia and Minhal for their precious contribution.

Finally my heartfelt thanks to my dear husband and family for the cooperation and support towards the completion of this work.

A very special thanks to Shahid Mahmood for the support in composing and finalizing this word.

PROF. DR. IRAM KAHLID
DEPARTMENT OF POLITICAL SCIENCE
UNIVERSITY OF THE PUNJAB

DEDICATION

Dedicated to all the brave sons of the soil, who sacrifice their lives for the safety and security of us.

Contents

INTRODUCTION

CHANGING PATTERN OF PAKISTAN'S NATIONAL SECURITY POLICY

As we write these words, our world is held in a strategic flux. The vulnerable state of Pakistan has to confront the challenges of this fluid external environment. Pakistan is facing a complex security environment haunted by a mix of traditional and nontraditional threats, both hard and soft. Some are enduring in nature, others are of more recent origin. On the other hand, the transnational nature of violent conflicts is posing a dynamic style of threat to Pakistan.

Modern concept of national security is no longer traditional, as once presented by Morgenthau where states were the principle actors in the international politics. With the growth of non-state and transnational state actors and the technological advancements, national security became a comprehensive sphere of state activity where internal conditions of states gained influence on transnational and international events. In the post 9/11 period, "the growing activities of transnational non-state organizations that pursue religious-cum-political agendas by violent and terrorist means have added to the security problems of the states".

Issues such as terrorism, weapon proliferation, spread of epidemics, transnational crime, narcotics smuggling, illegal immigration, women and children trafficking, piracy, money laundering etc. now count as issues of national security. While the traditional military security is relatively simple in content, nontraditional security issues include economic, financial, ecological and environmental securities, as well as consider information security and energy resource security as chief objectives while constructing national security paradigms.

A few important questions

- What type of decision anxiety is there in Pakistan due to security transition?
- Which issues are more critical in nature?
- How can Pakistan counter these challenges?

Changing National Security Pattern

- The changing notion of national security after 9/11 incidents has introduced new challenges to the world. Now security is not something limited to the state itself ,

- The issue of terrorism, and transnational extremist organizations, has involved the whole world. No country can counter the challenge by itself, or in isolation
- The case of Pakistan is no different; in fact, it is a model case constructed by traditional and non-traditional challenges, alike.

Place in Literature

- In his book Pakistan on the Brink: the future of Pakistan, Afghanistan and the West, Ahmed Rashid says, that the military establishment's tendency to pursue a national security paradigm and their influence on the foreign policy of Pakistan has placed the country in an economic crisis, damaged the social fabric and has contributed to the proliferation of religious extremism. He suggests that Pakistan look towards Turkey's successful mixture of economic growth, diplomatic sophistication and political stability. He urges that Pakistan allows a regional solution to the Afghanistan issue and focus on her own security through a comprehensive approach.
- Dr. Ahmad Faruqui in "Rethinking the National Security of Pakistan; the price of strategic myopia" makes a strong case for a thorough revision of Pakistan's national security policy and urges the policy makers to devise a comprehensive approach to national security that includes both the traditional and non-traditional security concerns. He analyses Pakistan's key issues such as political leadership, socioeconomic development and foreign policy with empirical economic and security data, and provides an expanded view of national security that involves social cohesion, political strength, economic development, diplomatic support and military readiness. The author suggests that Pakistan can reform its national security policy by integrating the military and non-military determinants in there.
- Alexander Burns in "Securitization in China and Pakistan's Frontier Regions: A comparative account of regional security discourses" makes a comparative analysis of Pakistan and China's national security discourses and states that "Pakistan has a highly complex set of securitization characteristics, something which stems from the unique relationship that exists between the military and the government as well as its lack of economic development. This has led to a peculiar fusion of the state, national and Islam in a way that

promotes a highly volatile and contested security discourse." He argues that the securitization theory cannot be properly applied to Pakistan or China because these states lack a democratic structure while the securitization process cannot be analysed in such an environment. Since securitization is a process that moves a politicized issue to be securitized, it requires a general run of affairs through "normal politics" under democratic rules until the issue becomes securitized and dealt as an emergency issue or "special politics". In an environment that lacks democratic rules of regulating political, economic and societal issues, securitization indeed becomes very difficult to function.

- The defense analyst Dr. Hasan Askari Rizvi Political in "National Security Council: A debate on institution and processes for decision-making on security issues" analyses the role of NSC in a number of democracies: the US, UK, India, Iran, Israel, Turkey, and Pakistan. He explains that in the established democracies, the military plays a marginal role in the last phases of the policy making process, while the experience of Pakistan and Turkey shows that the existence of the National Security Council (NSC) is "no guarantee that the military will not use other means to pursue its agendas in political and other domains and that it would not defy the constitution and democracy."

Pakistan's official perspective

- "An attempt to tackle domestic security challenges should be the main objectives for foreign policy and national security respectively," Sartaj Aziz, Special Adviser to the Prime Minister on National Security and Foreign Policy.
- "If we have a peaceful neighborhood, a very substantial part of our security dilemma gets sorted out," he emphasized. "Sorting out our relations with India is a top priority."
- After regional peace, he said, Pakistan needs to reset its relationship with the US after 2014 in a new "lasting" framework, one that goes beyond Afghanistan and in which the world superpower does not view Pakistan through the Afghan lens.
- According to me "National Security comes not from outside, it is not only the issue of power, actually it is based on internal strength of the state and society"

Emerging challenges to Pakistan

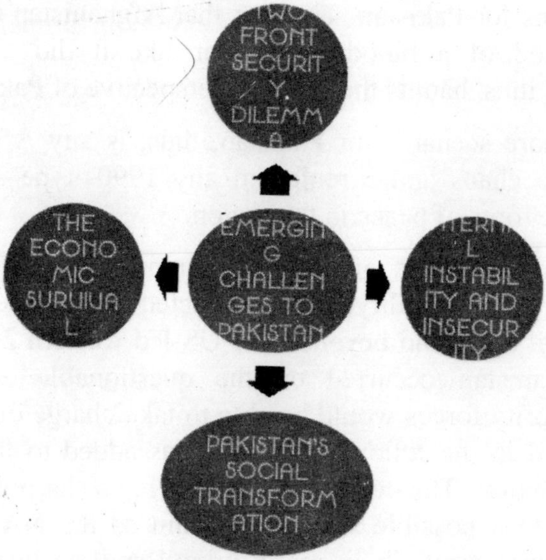

i) Two Front Threat/security Dilemma

- The traditional rivalry with India is still there but the threat on western border is posing new challenges in the shape of contiguous tensions with Afghanistan, that easily overflow the official borders in the form of extremists and their monetary or armed support.
- Pakistan faces several strategic and security challenges, and while atmospheric as well as the political and business environment has improved, the strategic relationship remains fraught. This is because of the unresolved state of disputes, including Kashmir, and India's persistence in continuing the build of conventional and strategic arms.
- The enunciation and operationalization of "proactive" military doctrines by India aimed at engaging in limited conventional war below the nuclear threshold is a further destabilising factor in the troubled strategic picture of Pakistan.
- The situation in Afghanistan has also posed enduring strategic dilemmas for Pakistan. Islamabad's security anxieties are being heightened by the uncertainties born as a result of the looming 2014 transition when most NATO combat forces departed Afghanistan. Having suffered from the blowback of three decades of strife and war

in Afghanistan, continued instability on the western front has serious repercussions for Pakistan. The fear that Afghanistan might slip back into a period of a bloody civil war like it did after the Soviet withdrawal, thus, haunts the security perspective of Pakistan.

- The nightmare scenario for Pakistan, thus, is any Afghan post-2014 descent into chaos and a replay of any 1990s-type scenario, which struck at the roots of peace in the region.
- Effective management of the 2014 transition is thus critical for Pakistan's internal stability, already affected by the spread of militancy into its tribal areas and beyond. The US-led western 2014 withdrawal from Afghanistan occurred on the questionable assumption that Afghan security forces would be able to take charge of security. What has followed in the following months has added to the uncertainties over that premise. This leaves Pakistan facing a sharp dilemma of how to respond to a possible security vacuum as the NATO withdrawal proceeds. Fears grow about more turmoil spilling over into its tribal areas and another influx of Afghan refugees to add to around 3 million already present in the country, both legally and illegally.
- The sanctuaries that Pakistani Taliban militants have found in Kunar and Nuristan, from where they conduct cross border attacks are a grim signpost to a heightening threat if the security vacuum widens in Afghanistan.
- More turmoil in Afghanistan will aggravate Pakistan's internal security when the threat from militancy from a syndicate of violent groups is far from being overcome. Violent extremism remains a daunting challenge and will preoccupy the country's army and law enforcement agencies and dominate the country's threat perceptions for years to come.
- Pakistan and Afghanistan share an immense border stretching 1510 miles (2430 km) along the southern and eastern edges of Afghanistan.[1] The Afghan provinces of Badakhshan, Nurestan, Konar, Nangarhar, Paktiya, Khost, Paktika, Zabul, Kandahar, Helmand, and Nimruz are all adjacent to Pakistan's border.
- Ethnic Pashtuns populate the area along the border. The frontier passes through varying terrain, with sandy deserts in the south and rugged mountains in the east.

- Major border crossings between the two countries are in Torkham, between Peshawar and Jalalabad and inSpinboldak, between Kandahar and Quetta. The border between the two countries was determined in 1893 in an agreement between the Afghan Emir Abdur Rahman Khan and the British Government of India. Since the creation of Pakistan in 1947, however, subsequent Afghan governments have not accepted the "Durand Line" as the boundary between the two countries. Kabul considers the dispute unresolved.

Several factors have coalesced to make the Pak-Afghan Border hard to guard:

a) Geography, as the area is too large to police properly; on the official border crossings, and along the line, law enforcement agencies have long aided or closed their eyes to the problematic cross-border traffic, of humans, drugs and terrorists

b) Since the Afghan Jihad against the Soviet Union, many have mastered the art of crossing the border without detection by authorities on at least the Afghan side of it

c) At many areas along the Durand Line, people from the same tribes live on both sides of the line, and move back and forth without much regard for the international boundary

Internal instability – insecurity/fear

- Internally Pakistan is at the lowest ebb security wise
- Terrorism, extremism, law and order situation are the major challenges confronting political stability
- Pakistan has a grave challenge in the shape of intra-ethnic conflict occurring across different areas. These divergent interests have plunged Pakistan into a whirlwind violence and chaos over the years
- Ethnicity is mixed with sectarianism, which is manipulated politically as well as economically. Karachi and Baluchistan are main examples.
- The inter-personal, inter- societal and inter-ethnic clashes have grown violent and now shape a severe threat to Pakistan's internal stability and security

- Killings in the name of ethnic and religious sanctity have become a routine matter in Pakistan
- The issue is so complex that the identification of enemy has turned into the main challenge for state
- The multidimensional and multilayered problems has gravely damaged the social fabric of Pakistan
- Internal disharmony and fragmentation has created a diverse set of problem for social cohesion to be possible

Pakistan's social transformation – from a modern to a reactionary society

- Pakistan, as a state, has to address multiple concerns regarding fault lines within its society. The chaos in the society is based on its ironic transformation from a moderate to a reactionary one.
- The sub-continent culture historically presented the picture of religious tolerance and accommodation. The Muslims ruled the majority and Pakistan inherited these values and practiced for decades. But the politics of experimentation and personal agendas of religious and political leadership destroyed the formation of Pakistani society.
- Now that the society no longer enjoys trust based relationship with state, the fragmentation within society the division has sharpened, to such an extent that it has created an actual threat to the national security.
- In the name of Islamization of the society and culture, certain cultural and social values, activities and norms have been discouraged or suppressed. Islamization has been misused as a political cause, pursued by the state as well as some local groups and organizations.
- In the process, some of the traditional sociocultural values, norms and activities, which long promoted and supported peace and harmony in society, were also targeted.
- In a way, the efforts toward Islamization on part of the state and non-state actors distorted the culture in such a way that it became somewhat accommodative for radical ideologies and mala fide practices in the name of religion and cultural integrity.

Factors behind transformation

a) Language problem: Despite the fact that people of Pakistan do recite Quran frequently, they largely do not understand Arabic; hence, they are dependent on the interpretation by their religious elites.

b) Sectarian approach: Teachings by religious preachers are mainly concerned first with their respective sects and vested interests and only after that with Islam, if at all. Driven by the primary concerns for their sects and personal benefits, they tend to preach a very selective interpretation of the Quran and Hadith.

c) Political use of religion: Religion has frequently been used for political ends. Politicians and ruler shave seldom missed an opportunity to exploit religion and the religious-minded communities in order to serve their vested political interests. "Politically motivated interpretations of Islam" are made for "opportunistic reasons". Due to this transformation people are being intolerant, both in their attitudes and beliefs.

iv) The economic survival:

- Economy refers to any arrangement where production, distribution, exchange, and consumption of goods and services take place.
- The economic performance of Pakistan has been substandard in recent years.
- Chronic power outages, violent crimes, and poor governance all hampered growth and contributed to falling foreign investment in Pakistan.
- Issues like poverty, unemployment, lack of proper health care, illiteracy and lack of justice are considered to be some of the major causes of militancy in Pakistan and elsewhere.
- Due to widespread unrest and political uncertainty in Afghanistan, large quantities of our food items/commodities are smuggled to Afghanistan, which ultimately leads to acute food grain scarcity within our own country.
- According to Pakistan's Finance Ministry, Pakistan suffered a loss of Rs. 2080 billion in the war against terror from 2004 to 2009, which severely crippled the national economic development

- Higher expenditures are invested on defense and armed forces, both military and civilian
- Dislocation of economic activities around the country has been a major problem.
- Decline in foreign and local investment has discouraged economic growth.
- High inflation rates are a death sentence for economic progress in Pakistan.
- During the last 10 years the direct and indirect cost of war on terror incurred by Pakistan amounted to $ 67.93 billion.
- Losses of countless precious human lives to crime and extremism have affected economy, and the country at large, more than anything else.

The 9/11 incident shook the international policy trends and reintroduced the new concept of external security. History took a whirl true after the US began the extensive global war on terrorism, by invading in Afghanistan. Since the attack, Pakistan's external, regional and bilateral security challenge have altered drastically; from the transnational, conventional to non-conventional fronts.

As far as the matter of internal security is concerned, the local drug cartels, criminal gangs and human traffickers as well as material black marketers have merged hands with terrorists for economic gains. Terrorism and local street crimes are the key trials with congregations of threats enablers, catalysts and multipliers which are; Feudalism, hereditary Political feasibilities, bad governance, zero to none accountability mechanism, fanaticism, communalism, weaponization, military's involvement in state affairs and weak political system.

Besides the aforementioned security, existential, economic, and political challenges, Pakistan also faces psychological, social and diplomatic challenges on the dynamic intangible battlefront.

Communal injustice, unequal distribution of wealth and resources, ethnic rift, sectarian split, high corruption, illiteracy and extremism are few problems which Pakistan has been dealing with, over the years. In the cyber domain, Pakistan has being long endangered to attacks and will endure to be highly prone to dangers, if no urgent policies are imposed to increase the cyber security. A pre-planned systematic cyber-attack at a critical time on our nuclear facility, data networks, telecommunication centers and electricity grid station can turn the worst case threat of exposing nuclear arsenal into the hand of radical terrorist groups, into a catastrophic reality. Moreover, the minds of public have become irrational as they are influenced and controlled by foreign aid and loans..

Even until today in the age of globalization, the horizon of Pakistan's politics is filled with feudal mindset, and authoritarianism governance, deficient of civilian institutions organizational structure and torn-society. If Pakistan's government and military doesn't learn lesson from their past experiences and continues in the same direction, the country would be soon experiencing more of disharmony in civilian-military relations, with resulting policies, which will jeopardize sovereignty and national honor and territorial integrity of the state at stake.

On the side part of non-conventional threats, One of the most serious and imminent challenge for Pakistan is water security. The high value strategic importance of this challenge is owed to numerous factors, predominantly the increasing pressure of population and subsequent increasing demand of water reserves, gigantic enlargement of tube well irrigation system, already protruding scarcity of clean water, reduced levels of rainfall caused by climate change, environmental change and the melting of glaciers.

Like any other sovereign country, external security has always been a principal concern for Pakistan. Two of the most recent and long going over looming security issues of Pakistan have recently developed salience in volatile environment of 21st century, therefore call for a new methodology of conflict resolution in the current regional setting; the ongoing armed conflict between India and Pakistan on the Line of Control (LoC) in Kashmir and the demarcated boundary line skirmishes between Kashmir and Pakistani territory in the Sialkot area of Pakistan; and the expected withdrawal of American led NATO troops from Afghanistan by the end of 2016.

The stalemate on the Kashmir dispute and the ongoing ever persistent tensions on the line of control are the mirror view of the biggest external security dilemma Pakistan has been facing since its independence. With India's growing power position and its emergence as an economic hub to grab the status of leader of South Asia, Indo-Israel nexus and the US's inclination towards India to contain and compete the overwhelmingly enormous power of china, Pak-India strategic equilibrium is expected to be altered.

Due to still hostile relations between Russia and USA, and competitive military buildups on both sides, the international actors are getting apprehensive of another cold war between Russia and America. In this background, Pakistan's dialogues and expectant arrangement with the Russians is another dynamic development. Pakistan has to be careful in dealing with the two rival powers of USA and Russia, with latter's aim to play a balanced role in Asia with counter-terrorism and economic plans, and Former's effort to secure her own national interests by swiping off terrorist group and access to natural minerals of South Asian continent.

The bordering-neighbor, troubled state of Afghanistan, has always been and still is a source of national threat for Pakistan. A high proportion of terrorists entered Pakistan from Afghanistan via boarder when US and NATO forces started a breakdown on them after 9/11 and they turned FATA areas of Pakistan into a terrorist training camp, a center for illegal arms and ammunition black market and a base for spilling instability.

The shifting geopolitical scenario after 9/11 and Arab spring, the declining natural resources, latest highly technological war methods begs Pakistan to introduce comprehensive changes in policy making techniques and political system. Moreover, it challenges and prompt us to plan ahead and be flexible enough to adapt with changing international trend in order to survive and prosper.

In addition to the interest of the regional powers, India, Iran, Russia, China and Saudi Arabia, U.S. post -2016 withdrawal from Afghanistan presents for a third world country like Pakistan with an open choice, whether she chooses to adopt the apparatus through which common interests can be successfully achieved and mutual beneficial partnerships can be attained and fostered or Pakistan chooses to remain alienated over conflicting interest.

SUGGESTIONS

- These external and internal challenges are mixed and Pakistan is trapped in the situation; there is no single strategy or way out to deal with the situation. Hence, a combined agenda is required.
- Pakistan must learn from the experiences of the states who have tackled the situation in a much better way.
- The state and society must be on one platform to find out a real solution.
- From individuals to groups, and from the society to the state, everyone has to take up the challenge at national level
- Responsible political leadership, having both necessary will and competence, is needed to tackle the challenges as a national policy.
- A dire need of time is to re-formulate the preferences regarding Pakistan's national security concerns.
- The leadership must first exactly define the problems and then show the political will to resolve the issues.
- Civil-military alliance has to be there to build a common narrative on national matters.
- Extremism gets a space when a country lacks cohesion and a sense of identity. Pakistan needs to seriously look into the challenge of nation-building so that the forces of extremism and radicalization are neutralized.

Conclusion

- One must keep in mind that only desires cannot resolve the issues; it is the practice which actually answers the difficult questions.
- The issues cannot be resolved with patch-work and isolated approaches.
- There must be a balance between defence and development to overcome the national security deficit.

CHAPTER NO - 1

NATIONAL SECURITY IN COMPARATIVE PERSPECTIVES

Introduction

As a nation challenged by a real existential dilemma at its very inception in 1947, the infant state of Pakistan automatically came to embrace an extremely tapered version of the doctrine of national security. However, it was the inherently flawed mechanism in the very policy-making environs of the country over the course of the following seven decades that continued to manufacture a mala fide narrative of it within the state, and the society, of Pakistan. Military security was equated with the national security as the primary narrative, as energies were predominantly invested at preserving Pakistan's territorial integrity and sovereignty by countering the historic Indian threat. All throughout, the principal decision makers belonged to the military establishment which naturally sought military omnipotence at the expense of other values that Pakistan had aspired to achieve as a part of its foundational ideology. Consequently, ideas such as democratization, nation-building, and development of sociopolitical, economic and moral infrastructure went to the backburner while the country plunged into an unending conquest of survival. So catastrophic was the result of this national crisis that the country lost its eastern half in 1971 and such profoundly hollow its promise was that the incomplete agenda of partition – the disputed State of Jammu and Kashmir – could not be completed till today.

Any country's military's principle task is to defend the state against external visible physical threats; it is not fit for the formation of a comprehensive national security policy, as "security concerns not only with the avoidance of loss, but also the prevention of blocked gain" (Hermann p.19). Nonetheless, Pakistan's military maintained its predominance over the state's domestic and foreign affairs, either through direct rule, or through aides such as the National Security Council during the years of impotent civil rule, causing further decay to the already retarded state institutions responsible for devising a comprehensive narrative on national security. A nondemocratic run of the political and diplomatic spheres of state activity to maneuver domestic and regional security challenges over the years, ironically, amplified the traditional security threats to the country and pushed the nontraditional security issues far behind in the political discourse. Thus, even after 68 years of independence, Pakistan maintains an ad-hoc national security policy at best, with a collage of defense, foreign, economic, diplomatic, and domestic procedures that simply do not add up to constitute a comprehensive strategy on national security. Pakistan never actually *nationalized* its security policy, nor did it exhaust its policy options to expand the scope of its state's security. This reflects Pakistan's limited understanding of the dynamic nature of security, and its traditional association of security with the status quo conditions. The quest of parity with India across a span of three generations only favoured to intensify the civil-military complex and subverted the possibility of the institutionalization of democracy in Pakistan. Thus, Pakistan transformed itself from a social welfare state envisioned by its founders to a state of national security, with a strong conviction of external but a poor assertion of internal sovereignty. Over the years the idea, that "national security has to go beyond a military-centric paradigm of defending physical borders to embrace a broader, more comprehensive approach... and that comprehensive security requires that all elements of national power – military, political, economic, diplomatic, cultural and informational – are deployed in a 'whole of government' approach so that multidimensional challenges are met by a holistic strategy" (Lodhi 2013) was lost deep in the grave of the colossal dream that had bore Pakistan.

In the post 9/11 international security environment, nontraditional security threats have dominated the national and regional security discourses around the world. All the responsible states of the world have focused on renovating their institutional and social infrastructure, devising policies aimed at maximizing their human resource, and creating greater opportunities for economic activity to cater the changing needs of their populations. Although Pakistan found itself confronted by a massive security challenge, both externally and internally, soon after the 9/11 attacks and the consequent invasion of Afghanistan by the US and its allies, it failed to address the panoramic threats to its national security due to a number of reasons stemming from its political history. This chapter studies the theoretical concept of national security, highlights the hurdles in the policy-making on issues of national security in Pakistan, and examines the national security policies of the US, the UK, France and China, in order to construct a working definition of national security in Pakistan's perspective, identify its real national security objectives, and suggest an institutional apparatus for attaining them with realistic hope.

Defining the Semantic Contours of National Security

There is an extensive debate across the academic circles regarding the concept of national security in the literature produced about it. The traditional or realist approach uses the state as the principal unit of analysis and defines national security in terms of the state's ability to deter an external threat to its national interests, thus, making the military the chief instrument of national security. For example, G. Kennan (1948) defines national security as "the continued ability of the country to pursue the development of its internal life without serious interference, or threat of interference, from foreign powers". Moreover, national security has been traditionally defined by academics and theorists to account for two major concepts: the threat being external, and the means to counter it being military. "When most people think of national security, they think of war. War and the preparation for war have been the constant focus of security practitioners since the birth of the modern nation-state... Since the military has been the traditional means of securing this power, national security has come to mean military might" (Chalecki 2013, p.2).

National security may be defined as "the ability to withstand aggression from abroad" (Luciani in Williams: 2008)

"National Security can most fruitfully be defined as the ability of a nation to protect its internal values from external threats." (Berkowitz and Bock 1965, pg. x)

This phenomenon dominated the early and peak Cold War years. A shift in the national security discourses was witnessed during the later years of the Cold War which took into account sociocultural and ideological factors. The Constructivist approach for example, focused on how norms, values, ideas, and beliefs of people influence their national security perspectives. It draws on the relation between the state's political culture and national security policy. With the demise of the Communist USSR and the shift of international power structure from bipolarity to unipolarity, the concept of national security became more wide-ranging and the mix of factors affecting national security changed. (Katzenstein 1996, p.1) Broader definitions of national security included the external and internal threats to the survival of a state and the threats to the conditions in which the state survives.

"A threat to national security is an action or consequence of events that (1) threatens drastically and over relatively brief span of time to degrade the quality of life for the inhabitants of a state, or (2) threatens significantly to narrow the range of policy choices available to the government of a state or to private, nongovernmental entities (persons, groups, corporations) within the state." (Ullman 1983: 133)

"The expectation of retaining and enhancing the ability to partake of highly regarded value outcomes free of obstructions." (Charles F. Hermann p.19)

In 1990, the nontraditional approach presented by Barry Buzan, Ole Weaver and Jaap de Vilde in their book "Security: a New Framework for Analysis", redefined the scope of national security. It included the study of non-state and transnational actors and transnational threats as national security concerns. It presented a framework for policy makers that could accommodate right about any threat as a national security issue, given the strict conditions the framework requires. Scholars of this school maintain that that the post Cold War international security landscape cannot be easily analyzed by the classical frameworks where states are the main actors.

"We argue against the view that the core of security studies is war and force. … Instead, we want to construct a more radical view of security studies by exploring threats to referent objects, and securitization of those threats, that are non military as well as military. … Threats and vulnerabilities can arise in many different areas, military and non military, but to count as security issues they have to meet strictly defined criteria that distinguish them from the normal run of the merely political." (Buzan et al. 1998, pp 4 - 5)

The authors argued that the contemporary world has witnessed the rising importance and influence of non-state and transnational actors and transnational threats such as terrorism and climate change. This approach provides a framework to integrate environmental, societal, economic, and political issues with conventional security to address a wide range of national security objectives. With the growing number of non-state and transnational state actors and the technological advancements, national security became a comprehensive sphere of state activity where internal conditions of states gained influence on transnational and international events. In the post 9/11 period, "the growing activities of trans-national non-state organizations that pursue religious-cum-political agendas by violent and terrorist means have added to the security problems of the states." (Rizvi 2012, p.9) Issues such as terrorism, weapon (and specially nuclear) proliferation, spread of epidemics, transnational crime, narcotics smuggling, illegal immigration, piracy, money laundering, human trafficking, etc. count as national security issues. While the traditional military security is relatively simple in content, nontraditional security issues include economic security, sociopolitical security, ecological and environmental security, information security, and energy security, in the list of objectives to be achieved while constructing a comprehensive national security policy.

National security thus is an intangible concept that can be assigned a variety of meanings according to a given security environment. In the words of Arnold Wolfers, national security is an "ambiguous symbol" – a set of concepts that are attended by great intellectual and political contestation, and which evolve over time. This belief is at the philosophical core of the national security discourses; however, all the schools of security studies agree the need to make specifications regarding the national security agenda in order to render a scientific usage to the elusive notion of the term. In the post 9/11 international security environment, where the threats faced by the states are no longer only external, formulating a purposive and inclusive national security policy has become an ever more important task for state administrations and policy-making bodies. Major powers such as the US, the UK, France and China have conscientiously put together their national security policies that define their role in the world, the threats they face, and the national objectives they aspire to achieve.

National Security Policies of Major Powers

According to the US Department of Defense (1914), National Security Policy is

"The art and science of developing and using political, economic, and psychological powers of a nation, together with its armed forces, during peace and war, to secure national objectives."

Identifying the national objectives and the means to achieve them is the main purpose of national security policy. Gen. Maxwell Taylor (1974) highlights the role of national security in the following words:

"The national valuables in this broad sense include current assets and national interests, as well as the sources of strength upon which our future as a nation depends. Some valuables are tangible and earthy; others are spiritual or intellectual. They range widely from political assets such as the Bill of Rights, our political institutions and international friendships, to many economic assets which radiate worldwide from a highly productive domestic economy supported by rich natural resources. It is the urgent need to protect valuables such as these which legitimizes and makes essential the role of national security."

In the post 9/11 era, the "homeland security" dimension has been integrated to the wider national security policy that takes into account domestic and foreign national interests of the state related to the state's survival, security, and development. This section highlights the national security of the US, UK, France, China, and the homeland security policy of India. The method followed for the presentation of these states' policies is to quote their priority goals for the national security agenda, which helps give the reader an image of each state's core national interests. This comparison is done in order to come up with a national and homeland security vision for Pakistan. The respective national security and defense strategies are discussed below:

I. US National Security Strategy 2010

In 2010, the White House highlighted "four enduring national interests" that were "inexorably intertwined" as the priority objectives that were to be pursued for the achievement of US national security:

- Security: The security of the United States, its citizens, and U.S. allies and partners.
- Prosperity: A strong, innovative, and growing U.S. economy in an open international economic system that promotes opportunity and prosperity.
- Values: Respect for universal values at home and around the world.
- International Order: An international order advanced by U.S. leadership that promotes peace, security, and opportunity through stronger cooperation to meet global challenges.

These objectives are comprehensive, as they include the major domestic and foreign national interests of the United States that are associated with the national security affairs. Security, for example, takes up issues of homeland security, management of the US forces abroad, coalitions and partnerships, diplomatic ventures, the US policies towards problems such as terrorism, weapons of mass destruction, nuclear proliferation, homeland security, etc. The objectives related to prosperity, on the other hand, address issues such as multiplying human and intellectual capital of the country, achieving balanced and sustainable growth, and enhancing science, technology, and innovation.

In *The National Security Policy Process: The National Security Council and Interagency System*, a report annually updated to present and review the evolving management process of the national security affairs, the authors discuss in detail the key departments involved in the national security policy process and their role in that process. They elaborate on the history of the National Security Council and the way the presidents of the United States systematically utilized this institution over the years to generate worthwhile policies on national security affairs, by bringing together the various departments of the state involved in its management. The report also discusses the National Security Staff, which includes the National Security Staff (NSS) and the Homeland Security Staff (HSS) and their various responsibilities that help make and implement an effective national security policy. It describes the current NSS organizational structure and processes, and defines the roles of key departments and agencies, including that of the National Security Staff. The report discusses how the interagency process is incorporating the relatively new organizational structures associated with homeland defense and homeland security.

II. A Strong Britain in an Age of Uncertainty: The National Security Strategy 2010

The National Security Strategy sets out two clear objectives:

- To ensure a secure and resilient UK by protecting our people, economy, infrastructure, territory and ways of life from all major risks that can affect us directly; and
- To shape a stable world, by acting to reduce the likelihood of risks affecting the UK or our interests overseas, and applying our instruments of power and influence to shape the global environment and tackle potential risks at source. It also sets out

in its National Security Risk Assessment a clear prioritization of those potential threats we face.

III. French White Paper on Defense and National Security 2008

The French White Paper on Defence and National Security, published on 2008, points out: "the national security strategy embraces both external and internal security, as well as the military, civilian, economic and diplomatic means. It must consider all the events, risks and threats which could endanger the life of the Nation". The goal of the national security strategy is to deal with the risks or threats which may affect the life of the Nation. The aims of the national security strategy is to

- Defend our population and territory
- Contribute to European and international security.
- Defend the values of the republican compact which binds together the French and their State: the principles of democracy, including individual and collective freedoms, respect of human dignity, solidarity, and justice.

IV. China's National Defense in 2010

The goals and tasks of China's national defense in the new era are defined as follows:

- **Safeguarding national sovereignty, security and interests of national development**. China's national defense is tasked to guard against and resist aggression, defend the security of China's lands, inland waters, territorial waters and airspace, safeguard its maritime rights and interests, and maintain its security interests in space, electromagnetic space and cyber space. It is also tasked to oppose and contain the separatist forces for "Taiwan independence," crack down on separatist forces for "East Turkistan independence" and "Tibet independence," and defend national sovereignty and territorial integrity. National defense is both subordinate to and in service of the country's development and security strategies. It safeguards this important period of strategic opportunities for national development. China implements the military strategy of active defense of the new era, adheres to the principles of independence and self-defense by the whole nation, strengthens the construction of its armed forces and that of its borders, territorial sea and territorial air defenses, and enhances national strategic capabilities. China consistently upholds the policy of no first use of nuclear weapons, adheres to a self-defensive nuclear strategy, and will never enter into a nuclear arms race with any other country.

- **Maintaining social harmony and stability**. The Chinese armed forces loyally follow the tenet of serving the people wholeheartedly, actively participate in and support national economic and social development, and safeguard national security and social stability, in accordance with the Chinese Constitution. Exercising to the fullest their advantageous conditions in human resources, equipment, technology and infrastructure, the Chinese Armed Forces contribute to the building of civilian infrastructure and other engineering construction projects, to poverty-alleviation initiatives, to improvements in people's livelihood, and to ecological and environmental conservation. They organize preparations for military operations other than war (MOOTW) in a scientific way, work out pre-designed strategic programs against non-traditional security threats, reinforce the building of specialized forces for emergency response, and enhance capabilities in counter-terrorism and stability maintenance emergency rescue. They resolutely undertake urgent, difficult, dangerous, and arduous tasks of emergency rescue and disaster relief, thereby securing lives and property of the people. Taking the maintenance of overall social stability as a critical task, the armed forces resolutely subdue all subversive and sabotage activities by hostile forces, as well as violent and terrorist activities. The Chinese Armed Forces carry on the glorious tradition of supporting the government and cherishing the people, strictly abide by state policies, laws and regulations and consolidate the unity between the military and the government and between the military and the people.
- **Accelerating the modernization of national defense and the armed forces**. Bearing in mind the primary task of accomplishing mechanization and attaining major progress in informatization by 2020, the People's Liberation Army (PLA) makes extensive use of its achievements in information technology, thus achieving the stepping up of the composite

and integrated development of mechanization and informatization. The state takes economic development and national defense building into simultaneous consideration, adopting a mode of integrated civilian-military development. It endeavors to establish and improve systems of weaponry and equipment research and manufacturing, military personnel training, and logistical support, that integrate military with civilian purposes and combine military efforts with civilian support. China vigorously and steadily advances reform of national defense and the armed forces, strengthens strategic planning and management, and endeavors to promote the scientific development of the national defense and armed forces.

- **Maintaining world peace and stability**. China consistently upholds the new security concepts of mutual trust, mutual benefit, equality and coordination, advocates the settlement of international disputes and regional flashpoint issues through peaceful means, opposes resort to the use or threat to use of force at will, opposes acts of aggression and expansion, and opposes hegemony and power politics in any form. China adheres to the concepts of openness, pragmatism and cooperation, expands its participation in international security cooperation, strengthens strategic coordination and consultation with major powers and neighboring countries, enhances military exchanges and cooperation with developing countries, and takes part in UN peace-keeping operations, maritime escort, international counter-terrorism cooperation, and disaster relief operations. In line with the principles of being just, reasonable, comprehensive and balanced, China stands for effective disarmament and arms control, and endeavors to maintain global strategic stability.

V- Homeland Security in India; White Paper

The way forward to strengthen the Homeland Security scenario in India would entail the following action steps, some of which have already been initiated by the Government of India:

- Strengthen the policy framework in order to fully realize India's vision for Homeland Security by encouraging greater public private participation, allowing access to and adoption of latest technologies and leveraging the growing defence sector specific competencies created within the country

- Further strengthening of the police and paramilitary forces and intelligence machinery at the Centre and in the States through better manpower, training, equipment and other related support
- Support greater private-public-partnerships and platforms for interactions between the Government and Corporate Institutions, creating a cohesive ecosystem
- Defining the minimum standards for contracts, equipments and training clearly to enable transparency in overall procurements
- Holistic expansion of defence offset policy under the Defence Procurement Procedure (DPP) to include equipment for the Homeland Security sector Enhanced critical infrastructure protection through increased physical security and better access control systems at vital installations by leveraging the latest technology
- Full and effective utilization of funds under the Police Modernization Scheme, Coastal Security Scheme, Backward Districts Initiative and other developmental schemes
- Improve the overall governance and delivery systems of public goods and services
- Citizen involvement initiatives aimed at encouraging citizens to partner in safeguarding the internal security environment of the country.

The following areas are identified as the priority homeland security areas in India:

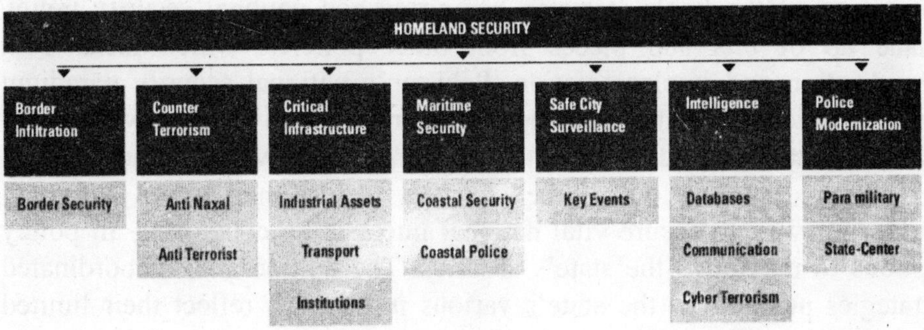

Homeland Security is an important dimension of the national security policy. This can be noticed in the national security strategies and policies of almost all the major powers in the world.

The Curious Case of Pakistan

Pakistan is one of the most vulnerable states of the world from the perspective of national security, particularly after the September 11 attacks. However, a quick look into Pakistan's national security paradigm reveals that Pakistan's political, economic, societal, and environmental issues have never been calculated into the state's national security policy as they are considered part of low politics while the decision makers with overarching influence on policy-making have often belonged to the state's military establishment. While discussing national security, one assumes that the state is well aware of its national interests in order to safeguard them through a particular national security policy. In Pakistan's case, national interest hardly goes beyond maintaining territorial integrity of Pakistan and achieving a sustainable military security in the face of foreign aggression, particularly the historic Indian threat. Since 1947, Pakistan has come along as a reactionary state responding to external threats usually only militarily instead of pursuing a coherent, constructive, consistent, dynamic and holistic security policy, mainly due to its narrow understanding of the concept of national security. Consequently, Pakistan focused on traditional measures for securing herself against foreign threats, many times security meaning simply military security. However, today's Pakistan faces many more threats than the single major and historic India threat; hence it needs to integrate nontraditional security measures to its national security policy to meet the security challenges posed by the regional security environment. Like discussed in the beginning of the book, Pakistan has an ad-hoc national security policy made up of bits and pieces from other policies, which lacks both coordination and implementation. Pakistan's national security paradigm remains defense-oriented; pursing the security objectives of powerful sub-national actors instead of the nation's interests as a whole. Policy-makers are required to employ all the elements of national power, not only military power, to secure vital national interests by exhausting all policy options to maximize the state's security. The ad-hoc and uncoordinated strategies pursued by the state's various institutions reflect their limited understanding of the dynamic quality of security and poor practical knowhow for pursuing a well-designed and harmonized grand strategy to meet national security challenges with clear and well-defined national security objectives. Pakistan requires a conscientious and consistent effort on the part of the policy-makers to identify critical high-value interests from secondary interests, evaluating the resources required for securing the identified priority national security objectives, and assessing the success of their efforts.

The following basis for a comprehensive national security policy is suggested:

I. Definition

Pakistan's national security policy can be defined as

"The National Security Policy of Pakistan is the framework of securing Pakistan's sovereignty and stability and a plan for advancing its national objectives, regionally and globally, through a policy of noninterference."

The above definition implies acquiring, retaining, and augmenting Pakistan's capacity to ensure the survival of its nationals in favorable conditions. It encompasses the dual functions of defending and developing the state of Pakistan using all the elements of national power in cohesion, in accordance with the international norms.

II. National Purpose

To emerge as a sovereign, independent, and a socially and economically developed state, in affirmation of the global and national values that ensure fundamental rights of existence for all its nationals.

III. National Security Objectives

National security is a function of the state's national purpose, which determines its national interests. To safeguard the state's national interest is the main objective of the state's national security policy. Thus, it is necessary to determine the following as Pakistan's vital national interests and permanent national security objectives:

1. *Security:* protection of its land and maritime borders and strategic assets.
2. *Stability:* national integration through institutional harmony and good governance.

3. *Prosperity:* advancing the economic well-being of the people Pakistan.
4. *Regional Order:* a regional order that is favorable to Pakistan's national security and does not undermine Pakistan's independent and sovereign status.

Based on these basic national interests, it is suggested that policy makers in Pakistan develops a "home first" approach to national security agenda, giving nontraditional security issues considerable priority over other national security objectives. The following table aims at highlighting some of the core national security objectives for the state of Pakistan.

	Global			**Correcting misperceptions regarding Pakistan's involvement in the internal affairs of other countries.**
	Domestic	**Nationhood building/ national integration**	**Betterment of law and order situation**	**Manufacture consensus on national security issues through institutional harmony**
				Steer the electronic and social media towards achieving national integration
Stability	**Regional**	**To secure a favorable regional order**		
	Global	**To reduce external debt and reduce dependence on external resources**		
		To sustain cooperation with friends and allies and pursuing an independent foreign policy		

				Maintaining smooth relations with the Muslim World. regional order
	Domestic	**Strengthening national capacity**		**energy and water security and debt reduction.**
				Preservation of basic freedoms and ensuring justice to the people of Pakistan.
Prosperity	**Regional**			**Play an active/influential role in economic organizations such as SAARC and even the Shanghai Corporation Organization.**
	Global		**To promote Pakistan as a potential Gulf state**	
			To enhance political and economic relations with Russia	
			To explore alternative markets for Pakistani products	
Regional Order	**Regional**		**Highlight the longstanding Kashmir dispute**	**Develop good trade relations with the neighbors; China, India, Iran, Afghanistan, Central Asia and the Gulf states.**
	Global	**To thrive as a viable state in the comity of**		**Active lobbying for securing national objectives in the changing international security environment.**

IV. <u>Institutional Apparatus and Policy Making Procedure</u>

A national security apparatus is administered by a blend of management practices and technical capabilities. One of the most important functional aspects of national security is the institutions' capacity and preparedness to handle national security issues, particularly in an environment where the state is faced with sudden unwarranted threats to national security, primarily in Pakistan's case, terrorism. Institutions that lack such capacity can actually themselves prove to be a threat to national security, since the security of the state relies heavily on the strength, efficacy, and harmony of its institutions. This is evident in Pakistan's case, hence, it requires a body of well-capacitated institutions to build consensus over the state's high-value national interests in order to put together and carry out an effective national security policy. It is also important to determine these institutions, which should include members from Ministry of Foreign Affairs, Ministry of Interior, and Ministry of Defense. The National Security Council is ideally meant to be a forum that provides consultation to the President and brings together the various departments of the state on matters of national security. However, the NSC as we have it today is dysfunctional as the civil-military relations have marred the usability of the NSC. The civil-military relations and institutional disharmony are discussed further in the next section.

V. <u>Hurdles in policy-making on national security issues:</u>

There are always a number of hurdles in the way of policy-making on national security issues in any country. These issues are not unique to Pakistan; all developing states face similar obstacles in their efforts to secure themselves. Some of these problems are:

1. Poor nationhood

A major challenge that Pakistan has to overcome in order to secure itself is the lack of national integration among, both, its individuals and institutions. The variety of identities assumed by the people of Pakistan based on creed, language, race, and class diversifies the nation and its resources in such a negative way that promotes divisions, separation, extremism etc. While poor nationhood is a hurdle to the implementation of a comprehensive national security policy, it also is prioritized as the chief challenge on the national security agenda since it has lead to the fragmentation of the society further in the after-shocks of 9/11 and intensified the nontraditional threats that arise from within the borders. Most of the internal terrorist threats that Pakistan faces today are due to the ill governance of people's political aspirations that arise from their identification with a particular creed or ethnicity. Overcoming this hurdle can ease the way for the realization of all other developmental goals.

According to a report on the *"National Security Strategy of Pakistan"* by The Institute of Strategic Studies, Islamabad (ISSI), Air Chief Marshal (Retd) Kalim Saadat led an In-House meeting where the agenda of the discussion was "Suggested National Security Strategy for Pakistan". The attendees concluded that the major challenges faced by Pakistan are poor nationhood, sectarianism, lack of economic well-being "due to absence of national integration, constitutional and educational challenges and agrarian dependence...Only a strong, democratic, dynamic and prosperous Pakistan can look after its people," and thus many of Pakistan's security challenges are internal. Without dealing with these issues Pakistan cannot properly secure itself against external threats either.

2. *Center-province relations*

Due to the poor sense of nationhood and the fact that the political parties in Pakistan show a sense of belonging to a particular province more obviously than toward the whole of the state, every time a civilian government came into power, little coordination was witnessed in the center-province relations. In fact, many allegations are thrown at the rulers at the center for favoring their own province over the others in order to secure their vote bank. Added to that is the problem of unequal availability and distribution of resources amongst the provinces and the fair management of such issues between the center-province relations. Even after the provinces are made independent, the capacity of the provinces to secure and develop themselves varies greatly, leading to a considerable level of dependence on the center. From a national security policy perspective, the ill-coordinated center-province relationship is a major hurdle in the way of effective implementation of a cogent national security policy.

3. *Civil-military interaction*

"National security issues cannot be left altogether with the top brass of the military. Similarly, the civilian political leadership cannot alone tackle them. Both need to consult each other... Military security is situated in political and diplomatic contexts that are taken care of by civilian political leadership. The latter also cultivates societal ownership and support for security policies. Such a societal support becomes more critical in the case of internal conflict, separatist movement and terrorist activity. National security is thus a joint or shared civilian-military enterprise. Appropriate institutions and processes are evolved for this purpose." (Rizvi, 2012, p.9)

Most of Pakistan's political history has witnessed unpleasant civil-military relations which have hollowed Pakistan's institutions. Due to the trust deficit between the military and civil leadership, there is a chronic lack of coordination and agreement over important national security issues and the way to go about addressing them. The National Security Council's weak role in the current scenario proves this claim to be true as the military is perceived to be seeking dominance over national security and foreign policy issues, and the civilian government feels insecure from such military involvement. Unless the state adapts to a complete democratic political system, it is difficult to generate consensus over national security issues amongst the concerned government departments and key state organs while holding each organ/department to carry out its own specified task without any interference in the others'. The defense analyst Dr. Hasan-Askari Rizvi Political in the *"National Security Council:A debate on institutions and processes for decision-making on security issues"* analyzes the role of NSC in a number of democracies: the US, UK, India, Iran, Israel, Turkey, and Pakistan. He explains that in the established democracies, the military plays a marginal role in the last phases of the policy making process, while the experience of Pakistan and Turkey shows that the existence of the National Security Council (NSC) is "no guarantee that the military will not use other means to pursue its agendas in political and other domains and that it would not defy the constitution and democracy."

4. *Lack of political will*

   Most politicians are obsessed with their narrow self-interests throughout their tenures instead of rectifying the state's security particularly from the nontraditional security threats. While the military has done a good job defending the state's borders, it has tended to define the state's national security in terms of defense mainly due to the vacuum left by the unwilling and many a times unfit political leaders, who could not neither define nor identify the nontraditional threats the states faces. While different political leaders aimed at specific developmental goals (in their home provinces) during their tenure, a comprehensive, well-calculated and foresighted policy to achieve developmental goals in all sectors across the geographical and political environs of the state was never devised by any political leadership. Due to poor governance and poor coordination, the state suffered heavily by the security environment created in the aftermath of the September 11 attacks. It is, therefore, imperative that politicians give up their old habits and develop sincere policies to relieve the nation from the major security challenges that have haunted the prospects of their entire lives, and construct a system of social justice for the people of Pakistan to allow the national security policy to function. Even if a proper national security policy is devised, it has no worth without a proper will to implement it and see it bear fruits, particularly when the entire character of the state stands distorted from social welfare to national security.

5. *Budget allocation*

   Another major hurdle that comes in the way of producing and implementing a well-architected national security policy is that there is hardly any separate budget set aside for the task. Since the concept of national security is not clearly understood by either the military or the civilian leadership in its true spirit, it has not been taken seriously enough to be given a reasonable share of the federal budget. Unless the formation and implementation of an intelligible national security becomes a priority task for the state, it is highly unlikely that Pakistan will achieve its national security objectives.

6. *Weak institutions and institutional disharmony*

A cursory look at Pakistan's political history reveals the level of disharmony in the state's institutions and the way it is reflected in the wider society. The task of consensus building over any issue is almost impossible due the suspicion and trust deficit in the state's institutions between each other. Moreover, the state institutions are sometimes so incredibly weak sometimes those individuals come to speak on their behalf. This magnifies the impact of the personality clashes between key players on the general run of the state affairs. Indeed, these institutions have themselves evolved into a threat to national security of Pakistan instead of acting as the safeguard; unless rectified they can have hazardous results. So far, the political and institutional hurdles have hampered the implementation of the policies related to security and development devised so far, and will continue to mar the national security of the state if left unimproved. The government should identify the causes of weakness in its institutions and eliminate them on urgent basis in order to neutralize the affect of personality clashes among leaders and enable the democratic process of governance.

It is a grave challenge to the national security policy-making that there is an absence of proper forums and institutions for forming coherent state policies that can connect different issue areas of security in a consistent national security policy. "Decisions on security issues remain ad hoc and disconnected, formulated and implemented in separate 'silos'. Policy is often reactive rather than coherent and consistent. Both civilians and the military eye much needed institutional improvisation with suspicion, believing that new structures will take power away from them." (Lodhi 2013) The inadequate institutional structure can make it even more difficult for states to deal with nontraditional security issues. Countries with weak institutions are vulnerable to nontraditional threats due to "insufficient institutional and physical preparedness. Since the 1990s, China has been frequently hit by nontraditional security threats such as the threat to economic security (East Asian financial crisis), hygiene security (SARS, poultry flu, AIDS, and so on), and environmental security threat (flood, sandstorm, drought, and so on)." (Burns, 2012, p. 46)

CONCLUSION

All of this follows the final idea, that national security is a state centric concept that is vague on its own, but could be given a well-defined structure by a state based on its:

- Geopolitical realities
- Historical experiences
- Resources
- Norms and Values
- Political culture
- National purpose
- Strategic environment

Each of these factors may not equally affect the national security objectives of the state. However, the above mentioned theoretical approaches use one or the other of these factors as the unit of analysis to develop a conceptual framework of national security. It is important to highlight that national security policy requires specification of objectives and the means to pursue them. It revolves around the processes of managing state security with well-defined security objectives that maybe rearranged in terms of priority, in order to score maximum objectives. Major powers such as the US, UK, and France have conscientiously put together their national security policies that define their role in the world, the threats they face, and the national objectives they aspire to achieve. Pakistan, however, has yet to formulate such a comprehensive national security policy that

- Is not only a defensive policy; instead it includes developmental/advancement aspects
- Is not confined to traditional threats, but also addresses external as well as internal threats
- Provides a firm, clear-cut stance on possible actions of retaliation in case the country's sovereignty is violated
- Highlights the vital, important, and significant national objectives

- Is devised by the representatives of all the institutions involved in advancing the state's national objectives.
- Provides the state's preferred methods/strategies of materializing those objectives with an assessment of the resources at the state's disposal.

Since the national security is a wide-ranging concept, Pakistan's political and military establishment along with general masses should devise Pakistan's strategic framework and produce a comprehensive national security strategy in a unified effort. The success of the political and military leadership in actually securing Pakistan lies in their collaboration in securitizing the nontraditional security threats and evolving relevant institutions for strategizing and implementing policies to meet that end. Pakistan's post 9/11 security environment demands a revision of the narrow outlook and faulty understanding of national security by the decision makers.

Summary

The art of agenda-setting is essential to national security. A well-planned and precisely calculated national security agenda can maximize state security. Pakistan always had an impromptu national security policy made up of a n incongruous mixture of other policies that lack sophistication and application.One of the most important functional aspects of national security is the institutions' capacity and preparedness to handle national security issues. Pakistan needs a capable national security apparatus that can serve the multifaceted objectives of national security. Policy-makers in Pakistan should make a conscientious and consistent effort to identify critical high-value interests from secondary interests, evaluate the resources required for securing the identified priority national security objectives, and assess the success of their efforts. It is suggested that Pakistan develops a "home first" approach to national security agenda, giving nontraditional security issues considerable priority over other national security objectives. The national security policy should be devised by the representatives of all the institutions involved in advancing the state's national objectives, particularly the MoF, MoI, and MoD.

References

1. Altaf, Beenish. (2013) *National Security strategy for Pakistan*. Retrieved from
2. http://paksoldiers.com/2013/02/26/pakistans-national-strategic-policy/
3. Baldwin, David A. (1997). The Concept of Security. *Review of International Studies,* Vol. 23. pp. 5–26.
4. Burns, A. (2012) *Securitization in China and Pakistan's Frontier Regions: A Comparative Account of Regional Security Discourses.* Lund University Graduate School Department of Political Science. Sweden.
5. Buzan, Barry, & Weaver, Ole, & Wilde, J. (1998) *Security A New Framework for Analysis.* Lynne Rienner Publishers Inc.: London.
6. Chalecki, Elizabeth L. (2013) *Environmental Security*. ABC-CLIO: Baltimore, MD, USA.
7. Chema, Pervez Iqbal. (2006) *The concept of comprehensive security*. The Post. Retrieved from
8. http://ipripak.org/articles/newspapers/theconcept.shtml
9. Faruqui, A. (2003) Rethinking the National Security of Pakistan: the Price of Strategic Myopia. Ashgate Publishing Company: Surrey, UK.
10. Herman, Charles F. (1982) *Defining National Security*. John Hopkins University Press: Baltimore, MD, USA.
11. Hafeez, Nasir. (2012) Evolution of national security structures in Pakistan.
12. Hussein, Syed Javed. (2013) *Towards a Comprehensive National Security Policy*. The Nation. Retrieved from:
13. http://www.nation.com.pk/columns/23-Jul-2013/towards-comprehensive-security
14. ISSI Report. (2011) National Security Strategy for Pakistan Institute of strategic studies. Islamabad.
15. Lodhi, Maleeha. (2013) *Wanted: A National Security Strategy.* The News. Retreived from:
16. http://www.thenews.com.pk/Todays-News-9-175900-Wanted-a-national-security-strategy
17. Rizvi, Hasan A. (2012) National Security Council:A debate on institutions and processes for decision-making on security issues. *Civil-Military Relations.* PILDAT: Pakistan.
18. Katzenstein, Peter J. (1996) *The Culture of National Security: Norms and Identity in World Politics.* Columbia University Press: NY
19. Report. (2011). The Evolution of Concept of Security. IEEE, *Spanish Institute of Strategic Studies.* Madrid, Spain.
20. Report. (2011) The National Security Policy Process: The National Security Council and Interagency System. Downloaded from http://www.ndu.edu/icaf/
21. Taureck, Rita. (2006) Securitization Theory and Securitization Studies. *Journal of International Relations and Development.* (9) pp. 53-61
22. Nieto, Pedro Rivas & Marcos, Roberto Gelado. (2012) Political realism and national security. *Public Administration Research,* Vol. 1, No. 1. pp. 24-32.

23. Lantis, Jeffery S. (2002). Cultural theory and national security. Blackwell Publishing, Malden:UK. p. 1-3

24. White Paper. (2008) The French White Paper on defence and national security.

25. White Paper. (2010) A strong Britain in an age of uncertainty: the national security strategy.

26. White Paper. (2010) National Security Strategy. The White House, Washington, USA.

27. White Paper. (2010) Homeland Security in India. ASSOCHAM.

28. White Paper. (2010) China's National Defense. Information Office of the State Council of the People's Republic of China

29. Downloaded from the official Chinese Government English translation at URL:

30. http://www.china.org.cn/government/whitepaper/node_7114675.htm on 31 March 2011

31. Wolfers, Arnold. (1952). "National Security" as an Ambiguous Symbol. *Political Science Quarterly*, Vol. 67, No. 4. pp. 481-502. Retrieved from:

32. http://links.jstor.org/sici?sici=00323195%28195212%2967%3A4%3C481%3A%22SAAAS%3E2.0.CO%3B2-R

CHAPTER NO. - 2

NATIONAL SECURITY

Introduction

National Security of any state is essentially a social manifestation and an equivocal perception. It is a comprehensive idea than to gain more power. The major aspect of national security is preservation of publics' homeland including their territories. So, national security means that how a state can preserve its prestige as well as economic interests and national values at international level. In this context, security relates to any state's desire to survive from internal as well as external threats positively.

Security paradigm is more complicated and incomprehensible phenomenon due to consistent set of challenges instigated by strategic principles of states. The external environment of any country mostly creates security patterns inside and outside the country. But, the internal security lapses primarily links with state's economic potentials. No country in the world can make progress itself until it has clear planning for future in the light of present circumstances. Such pre-cautionary measures minimizes the surety threats.

Pakistan as being the front line state in the global war against terrorism, its strategic security apparatus is far from satisfactory. Due to this global war, its economy is sinking, compelling it to depend on foreign countries and international organizations for economic support. This economically dependence has resulted into critical loss of economic self-reliance.

In political terms, Pakistan has the problem of weak governance which have non-democratic environment in the country. In addition, unfortunately Pakistan has dictatorial culture within the political parties. The decision makers of policy formulation in the country are making policy decisions without considering the institutional outputs which are resulted in threat multiplier in the long term security patterns.

Most importantly, the distressed areas of country such as Federally Administered Tribal Areas (FATA), Baluchistan and Karachi, security measures apprehensively in all cases are unreliable and fragile. In fact, Pakistan is still not succeed to block the aiding of sectarianism and extremism within country that were historically initiated in the course of late 1970s. Pakistan is still giving heavy pay back of those previous flawed domestic policies.

Internally, Pakistan is also plunged into a number of risks related to domestic emergencies. Its national security is also connected with energy crisis and water scarcity in the country and the lapses of long run planning to solve these issues positively. Pakistan is also facing the exhaustive population growth rates and a flow of urbanization which is also not suitable for a country having an agrarian economy. Furthermore, natural calamity is a phenomenon necessarily linked with country's national security patterns and has potentially security apprehensions for Pakistani population residing alongside the coastline.

So, recently Pakistan is portraying a bleak picture in which sense of insecurity prevails everywhere. There is also an alarming conditions about Pakistan's economic, societal and political lapses and the issue of radicalization and weaponization within the society.

So, the main objective of this research work is to logically explore the gap between threat perception and policy options because with the passage of time, the security threat to Pakistan national security apparatus is mounting gradually and also to investigate the lapses in policy options that surely could be the reason of increasing threat to national security of Pakistan.

Therefore, the current study is being conducted to confer about another matter of concern that although, Pakistan security institutions are doing their best jobs to curb the national security threats positively but Pakistan is basically fail to take pre-cautionary measures to obstruct such national security threats to arise accordingly and the fault lines in national security paradigms will also be analyzed that might be the hurdle in curbing the threats to its national security.

And lastly the research work will find out the best policy options that may boost up the security measures to Pakistan national security as Pakistan can flourish itself to the arena of global world positively.

Review of the Literature

Manzar Zaidi narrates in his book titled "Insight on Insecurity in Pakistan" that terrorism has become a global phenomenon and the global world is doing their best arrangements to deal the terrorist elements. This book also tells us what the terrorism is and what are the main objectives of these extremists which they want to achieve from Pakistan. It also describes the actual drivers of these activities. The writers has also given the bulk of knowledge about these elements as how do they operate in the country, how do they identify each other in other states and join hands in doing terrorist activities. The book has also logically narrated the factors involve in cementing them with each other and how states particularly like Pakistan can engage these gurrellas. So, the book has exhaustive knowledge about said questions and is much helpful to deal these terrorist elements by Pakistan government positively. (Zaidi, 2012)

Another book "The Future of Pakistan" led by famous scholar Stephen P. Cohen with some well-known contributors, has given the comprehensive information about Pakistan's current political, social, economic and religious scenario in terms of its national security paradigms. The book instigates with Cohen's comprehensive summary of Pakistan particularly in the terms of today geo-politics and international economics. All the contributors have logically highlighted the Pakistan's security apparatus and also different security threats such as terrorism, weaponization, sectarianism and sluggish economy and its aftereffects at national, regional and global level. "*The Future of Pakistan"* assists us to comprehend the country's recent security conditions, the pertinent actors and their inspirations, and different outcomes they might induce, and what it all means for Pakistan, India, the United States, and the whole world. (Cohan, 2011)

Stephen Philip Cohen, in **"The Idea of Pakistan"** has critically analyzed the predicaments presently challenged to the Pakistan. Cohen portraits the real image of Pakistan which has its own identity, logic and vision. The book narrates the complexity of the country that is in continuous position to equalize its multi-based identities and objectives with the changing nature of patterns of the international and regional politics and analyzing these patterns, Cohen adeptly narrates the trends of the country in past, present and future respectively. In his analysis, the writer successfully has addressed the economic, educational and demographic structure of Pakistan and how it can be maintained into stability and development after solving the constraints. Consequently, Cohan has given the plausible futures for Pakistan. (Cohan, 2004)

Zahid Hussain analyzes in **"Frontline Pakistan: The Struggle with Militant Islam"** that Pakistan was in much trouble after the devastating event of the 9/11 attacks and forcibly accepted the status of frontline state in the waging of war against terrorism and changed its previous policy of supporting Taliban to chasing them with US coalition forces. Likewise, he discusses the role of madrassas in the production of extremist elements in the country and critically analyzes emergence of sectarian violence in Pakistan. He spotlights the how militancy is a horrible threat to the Pakistan's security. (Hussain, 2007)

Pakistan: Democracy, Development and Security Issues" by **Veena Kukreja, M. P. Singh** has given the exhaustive knowledge the antagonistic conceptions of democracy, development, and national security in Pakistan. The book is about the external as well as internal challenges facing by Pakistan and how democracy has prevailed in Pakistan. (Kukreja&Singh, 2005)

The book, "Pakistan-Beyond the Crisis State" by Maleeha Lodhi has successfully explored the fault lines of Pakistan and how to overcome these lay outs. Pakistan's internal and external challenges such as political, economic, social, foreign policy and governance are logically analyzed in detail and particularly external role and their vital interests that influenced Pakistan in the formulation of its fortune. Maleeha Lodhi and other contributors stress that the Pakistani nation has the capability to bring her country on the way to progress and stability but bold and innovative reforms will be required to get proposed results. (Lodhi, 2012)

The above mentioned literature has given the detail analysis of the Pakistan national security patterns and threat perception to country's security apparatus respectively. But the main objective of this research work is to logically explore the gap between threat perception and policy options because with the passage of time, the security threat to Pakistan national security apparatus is mounting gradually and also to investigate the lapses in policy options that surely could be the reason of increasing threat to national security of Pakistan.

The study will also confer about another matter of concern that although Pakistan security institutions are doing their best jobs to curb the national security threats positively but Pakistan is basically fail to take pre-cautionary measures to obstruct such national security threats to arise accordingly and it will also explore the fault lines in national security paradigms that might be the hurdle in curbing the threats to its national security.

And lastly the research work will find out the best policy options that may boost up the security measures to Pakistan national security as Pakistan can flourish itself to the arena of global world positively.

Research Questions

- Why there is a gap between threat perception and patterns of policy options to curb these threats, as with the passage of time these threats to Pakistan national security is enhancing gradually?
- What are the lapses in policy options that could surely be the reason of mounting threats to national security of Pakistan?
- Why Pakistan is fail to take pre-cautionary measures to slab such threats to its national security positively and what are the fault lines in national security paradigms that might be the hurdle in curbing these threats?

Definition of National Security

Word "security" is a controversial concept in the arena of social sciences. As there is a great difference of opinion upon the meaning of word security. The theoretical concept on the basic question as to what security means and denotes and how it gradually calculates in global politics. So, theoretically it remains a controversial phenomenon between the perceptions of traditionalists and non-traditionalists. (Baylis, 1997)

So, national security can be defined in two ways;

- Traditional Security
- Human Security

Traditional Security:

It is an old, narrow and traditional model of national security which based on the notion of *"survival of fittest against external adversaries through military might"*. It has remained the prominent feature in the national policy making process of developing countries especially in states having dictatorships. Security simply is safety from harm or security against anything bad. It is one of the basic requirement of human beings. American psychologist, Abraham Maslow, described safety subsequent to the biological needs of human beings in his legendary "Maslow's Hierarchy of Needs Theory". After the fulfillment of basic necessities such as eating, drinking, sleeping etc., there is need of safety and security. So, we may apprehend the significance of security.

Historically, Security studies has remained the conspicuous issue in the arena of international relations. Traditionalist's school of thoughts is state-centric in conventions about security and international politics. Security can be determined in military means and safeguarding territorial reliability, sovereignty and independence from external extortions at international level.

In the words of Steven E. Miller,

The end of the Cold War meant the close of an era in the world and in the field, but the new era with its ongoing blight of political violence...War and violence remain with us. The intentional use of organized force for political ends remains a common phenomenon, one with large political, social, human, and economic costs and consequences. (Miller, Summer 2001)

According to Realist school of thoughts, state is the most imperative factor leading to idea of security and tries its best to enhance its power. The menace to national security comes mainly from other states. In order to preserve self-defense, the states have to build up their military power. Although, Realist school of thoughts are in favor of other necessary powers like wealth and geo-political position but military power is sole factor for national security.2 such kind of analysis leads to the enhancement of national armies and gigantic defense budget, especially during the Cold War. The whole world was frightened from use of nuclear arsenals during the Cold War. Consequently, security was intensely rooted in the realist patterns as threat perceptions and use of powers related to inter-state conflict or war. (Syed, Summer 2014)

Human Security:

The discussions on the conception and structure of national security started in 1983 as the Cold War was at its peak. Richard Ullman questioned the only military impression of security by analyzing that it gave the wrong interpretation of the reality as the states concentrates only on military threat at the cost of overlooking some other real and critical threats. Such false interpretation of national security leads to militarization of international relations causing international insecurity. (Smith, 1999)

In 1989, Jessica Tuchman Mathews gave new definition of national security by incorporating resources, environmental and demographic questions.

In the post-Cold War era, the growing of ethnic identities and cultural variances gave new directions to the nature of conflict from inter-state to intra-state. The concurrent progression of globalization has also developed the concept of human security. Globalization is a gradual process which had zoomed up in the 1990's owing to three dynamic forces as economic liberalization, political liberalization and speedy flow of information and communication technologies. But at the same time, new threats like climate change, food scarcity, terrorism, human trafficking has also emerged. Presently, 842 million people in the world, mostly from the developing counties are starving for food. Approximately seven million children die every year under the age of five just because of shortage of basic medical facilities, sanitary and drinking water facilities. (Syed, Summer 2014)

The nature of new threats such as global warming has also questioned the capability of states to deal these security threats and to develop a global approach towards these problems.

Famous scholar Barry Buzan explains security in terms of political, economic, societal, environmental and military sectors which envisions three level of security challenges as at sub-state, state and international level. (Ibid)

The concept of human security was firstly coined by Dr. Mehbub-ul-Haq, former Finance Minister of Pakistan in 1994 in the Human Development Report of United Nations Development Programme (UNDP). This report emphasized that common man wants security from threats of disease, starvation, unemployment, crimes, socio-political repression and environmental hazards. (United Development Human Development (UNDP) Report, 1994)

The human security concept was firstly anticipated for the "World Summit 1995 on Social Development". The core aspect of that summit was to concentrate on economic and social concerns. This concept is also imbedded with the Universal Declaration of Human Rights, Geneva Conventions and the United Nations Charter. (Tzifakis, 2011)

The UNDP Report of 1994 concentrated that human security is based on two aspects: firstly, security from chronic extortions like starvation, disease and repression and secondly, protection from unexpected and dangerous disruptions daily life.

The concept of human security is a great development in international relations and changed the patterns of security paradigms from the traditional concept to human security concept. It has changed the nature of national security from narrowly-defined militaristic concept to broader sense in terms of human being's welfare. Actually the concentration of security paradigms has shifted from states to general public.

Patterns of Pakistan National Security

Words "security" and "threat" are related to each other. But before doing any useful discussion upon security or threat perception, the entire scenario should be comprehended properly. So, the threats and sources of insecurity should be carefully analyzed before announcing any national security strategy of any state. (Musarrat, et, al., 2013)

Today, Pakistan is facing a complicated security landscape and severe challenges to peace and constancy. Due to internal security lapses and weak law and order situation, Pakistan is today being considered same as the countries like Iraq and Afghanistan. International media is interpreting it like *"the most dangerous place"* in the world. Challenges to peace and constancy originates from various array of sources and comprise traditional and non-traditional or asymmetrical threats. The basic threats to Pakistan national security are internal threats, but it has also been facing some persistent external challenges as well. So, the current research work is analyzing both patterns of threat external as well as internal to explore the Grand National Strategy for security to eradicate these threats positively. (Khan. February, 2014)

- **External Security Patterns**

  Pakistan has been confronting blend of external threats to its sovereignty and national integrity just after independence. Intermittent internal predicaments/disturbances further complicated the situation. As being sandwiched between India and Afghanistan, Pakistan's security insights have been essentially affected by its nature of relations with both countries. But most security apprehensions are from India.The devastating event of 9/11 have further complicated the situation. Most important external patterns of threats are such as;

- ## Indian Factor - A Primary Threat to Pakistan National Security

After independence, India and Pakistan have fought three major wars (1948, 1965, and 1971), observed two big border clashes (Runn of Kutchh in 1965, and Kargil in1999) faced three main crises (Bras stack in 1987, Kashmir in 1990, and troops confrontation in 2001-2), and had exchanged a lot of cross-firing at LOC. Both countries are facing a lot of reservations, mutual mistrust, and resentment towards each other. Undoubtedly, the Kashmir dispute isthe foremost bone of contention between the two countries. After independence, the state of Kashmir has faced severe ferocity and various types of predicaments. Kashmiri people have been constantly struggling and enthusiastically demanding independence.

The Indians are blaming Pakistan for backing such kind of slogans and refuses to admit the Kashmiri voices for independence. Resultantly, More than 100,000 Kashmiri persons have lost their lives. Apart from the Kashmir issue, Siachin, Sir Creek, and water crisis matters are also lighting the fire between two nuclear neighbors.

On the other side, India's increasing budget for defense shows its objectives in new regional security environments. Allocation of such heavy defensive budget is primarily Pakistan centric to enhance its conventional as well as ballistic capabilities to create a serious conventional imbalance in the region. The devastating sophistication of India's conventional military centered on the perception of "Cold Start" strategy, duly hooked by naval forces and strengthened by its second strike nuclear capability increases the mode of insecurity to Pakistan national security apparatus negatively. (Shoukat, August 3, 2009)

Although the process of Confidence Building Measures (CBM) was started in 2004 to normalize the relations between India and Pakistan. But the Mumbai attacks in India abruptly terminated the progression. So, such kind of events disturbed the restoration of the expedient dialogue process. In addition, the process of normalizing the relations has always been disturbed due to deficiency of meaningful momentum to resolve the disputes. Conclusively, it provides the room to Non-State Actors and extremists in both countries to take benefit of the situation from time to time. (Haq, April 18, 2013)

- **Afghanistan Issue - Un Predictable Western Front**

Another major external menace to Pakistan national security instigates from the existing developments in Afghanistan. The threat from Afghanistan side was arisen particularly after Soviet invasions in Afghanistan in 1979 and since then, Pak-Afghan relations have remained in many vicissitudes. (Cheema, September 2006)

But, 9/11 incident further complicated the situation. The some powerful elements within Afghan administration are in favor of cordial relations with India and for this purpose they do everything to create problematic environment for Pakistan from Afghan side. They are working at the behest of their external patrons. In fact, in strategic terms, the internal problems of Afghanistan including the current civil war cannot pose a grave menace to Pakistan's national security. But the threat of possible disbursement of political instability of Afghanistan into the adjoining provinces of Pakistan. The provinces of Khyber-Pakhtoonkhwa, FATA (Federally Administered Tribal Areas), and Baluchistan are enormously affected of such situations primarily because of the existence of dissident fundamentals. (Khan, 2011)

But the most problematic factor is that these terrorist elements are being encouraged and materially supported from India to threaten Pakistan national security and Afghanistan land is being used for this purpose. US Secretary of Defense, Senator Chuck Hagel remarked that India is financing the dissident elements to create security threats for Pakistan from Afghanistan.

Six issues have been negatively affecting Pak-Afghan relations rather consistently since the independence such as unrestrained reports about the Taliban's links with Pakistan, the contentious status of Durand Line, Indo-Afghan relations, the landlocked nature of Afghanistan, the existence of foreign forces (including American and NATO forces), and the undesired critical speeches of the Afghan high ups unnecessarily antagonizing Pakistan. Each of these issues has lit the fire in adverse relations periodically. Soon after the tragic events of 9/11, US attacked Afghanistan and further complicated the situation in Afghanistan.

In order to stabilize the country internally as well as externally,

Afghanistan has to join hands with Pakistan and should create massive and collective approach to defeat the common enemy. Pakistan is wanted to see a stable and peace Afghanistan as next door neighbor. In addition, the anticipated gas pipeline (TAPI) from Turkmenistan can only be secured if there is a stable Afghanistan. Because Afghan instability has not only blocked Pakistan's access to Central Asian States but badly affected upon Pakistan's potentials as an energy and trade corridor, it has also seriously affected stability in Pakistani border areas. Besides, the threat of expected instability and the likely occurrence of civil war within Afghanistan in the post ISAF (International Security Assistance Force) drawdown are constantly on the rise. Indeed, Pakistan does not want to see Afghanistan becoming safe havens for anti-Pakistan elements. (Cheema, July-September 2006)

- **US Hegemonic Behaviour - Controversial Benchmarking of Financial Aid**

Pakistani observations are that the US has frequently used and abused Pakistan for the maintenance of its vital interests. The analysts narrate about constant American infidelity in Sino-Indian war of 1962 in which it could force India to solve Kashmir issue and again in 1965 war of Indo-Pak in which US had stopped aiding Pakistan, Although Pakistan was its formal ally and once again US negative role in 1971 war resulted in the dismemberment of Pakistan. (Kux, 2001)

Then the stoppage of militaries supplies in the Pressler amendment by US and trust deficit behaviour of Americans in Afghan invasion were only the most recent examples of American betrayals. This approach remains currently with regard to the US-India nuclear agreement.

The most negative factor of using Pakistani land to plan to attack on Americans. An incident of successful attack at "Times Square" bombing in May 2010 which had its connection with Pakistan, led to strong Congressional reaction to punish Pakistan, or minimum to stop rewarding it. (Cohan, 2011)

In December 2010, there were in formations about the sending of Afghan "militia" groups inside Pakistan to attack militant groups operating in Afghanistan from Pakistani territory. But, most of the US invasion into Pakistan in the shape of drone strikes. Their targets in

Pakistan are al-Qaeda, the Afghan Taliban which US blamed of having safe havens in Pakistan or other guerrilla groups, such as the Haqqani network and Gulbuddin Hekmatyar, had basis in Pakistan but actually part of the Afghanistan political system. Most of the Pakistanis believes that radicalization of the state is basically due to US counterterrorism policies and military operations in Afghanistan. Therefore, most of Pakistanis consider US as great threat to Pakistan national security than India. (BBC, April 18, 2010)

The US has amplified its help to Pakistan conditionally on the basis of Pakistan's performance in combating the terrorism in FATA. Such hassle of "conditions" for the delivery of support, which pervades US strategy, irritates Pakistani community and causing all these steps to be fruitless, decreasing the confidence between US and Pakistan and mounting anti-US sentiments among Pakistani public. (Khan et, al, 2014)

- **Global terrorist Network - Al-Qaeda and 4th Generation War**

The Al-Qaeda is a worldwide movement of terrorist organizations, networks, and personnel. It is an organized organization which has many cells working in many countries. They are interconnected but only in the shape of one cause, if one person or group is caught in any country, the other members are not in danger. Counterterrorism forces of countries are unable to locate and investigate its specifications as its actual strength, historical structure, nature of organization etc. Estimations about its specification are often based upon rough calculation. These calçulations are inaccurate because the total number of operational camps is not definitely decided. (Jaspal, winter, 2010)

The subsequent grave menace to Pakistan's security and constancy narrates to asymmetrical or 4th Generation War (4GW). Pakistan is plunged at both levels as internal and external in this war. The activists of such kind of war are mostly locals but their logistical and financial support is carried out by foreigner patrons. The 'ungoverned places having weak law and order situations particularly in big cities, having already a bulk of population, provide them safe havens. It is not easy for satellite and investigation structure to locate them or to separate these extremists from ordinary citizens. (Zaidi, Feb. 22, 2011)

Al-Qaeda has safe havens at Af-Pak border areas and with the collaboration of local militant groups and organizations like Tehrik-e-Taliban Pakistan (TTP) is working to disrupt the political system, destabilize economic infrastructure and disintegrate social coordination within Pakistan. Ayman al-Zawahiri publicized,

"Muslims of Pakistan! You must now back the mujahedeen in Afghanistan with your persons, wealth, opinion and expertise, because the jihad in Afghanistan is the door to salvation for Afghanistan, Pakistan and the rest of the region. Die honorably in the fields of jihad". (Dawn, July 17, 207)

Pakistani President Pervaiz Musharraf rightly condemned these religious forces preaching for extremism in the world who used mosques for such purposes. Since 9/11, Pakistan has been actively playing the role of frontline state in the war against terrorism also paying a heavy losses. But the longevity of war has been enhancing extremism in FATA, Khyber Pakhtoonkhwa and Balochistan respectively which is being caused of bringing about socio-economic insecurity and political anarchy in Pakistan.

Recently, a military operation against terrorism is started in June 2014 at the tribal belt of North Waziristan in Pakistan. The proclamation of operation firstly came from the army rather than civil government. The spokesman from the army headquarter stated in a press release,

"On the direction of the government, the Armed Forces of Pakistan have launched a comprehensive operation against the foreign and local terrorists who are hiding in sanctuaries in North Waziristan. The operation has been named Zarb-e-Azab". (Khan, 2014) Spokesman declared that army was drawing no difference between "good" and "bad" Taliban unlike the past. As these terrorists are using Af-Pak tribal belt as their safe haven to operate to destabilize systems in Afghanistan and Pakistan. Taliban's reaction to operation Zarb-e-Azab was in anger and they said that they will attack on urban centers, government installation, foreigners in retaliation. Pakistani Taliban said in a statement a day after the operation began,

"We warn all foreign investors, airline companies and multilateral institutions to suspend all their affairs in Pakistan, and prepare to leave Pakistan". (Shah, 2014)

However, the recent operation in Bajaur, Mohmand, Orakzai, and South Waziristan agencies looks to have achieved considerable gains. But it would become crystal clear in the future about the outcomes. However, a comprehensive counterterrorism strategy is necessary for the safety and security of Pakistan.

- **Nuclearization of the Region – Deterrence factor**

Pakistan is one of those countries of the world who has nuclear capability. Pakistan got its nuclear capability from other countries in order to boost up its national security paradigm. Somewhere nuclear weapons are involved, war in the form of an organized battle between states applying the latest and most destructive weapons is hardly possible. In the case of Pakistan and India, the attainment of nuclear weapon is only to create deterrence in the region as one could not dominate other.

Dr. Samar Mubarak, one of the leaders of the Pakistani weapons design team said correctly in a statement, " that if it were not for Pakistan's nuclear deterrent, Pakistan would have not survived after Kargil, Indian parliament attack and Mumbai incident episodes". (The News (Lahore), May 31, 2009)

It possibly cannot be used for an organized war instead of any madman comes at helm of power at any side. But the great threat of nuclear theft particularly from any terrorist element or the conscious transfer of nuclear technology. Particularly from the westerns media, a lot of concerns upon the Pakistan's nuclear safety and security is being felt. These apprehensions are the theft of nuclear weapon by terrorist network, vulnerability of expected wartime scenario particularly during its movement for deployment. They are worried that domestic instability in Pakistan can loosen the government control over nuclear arsenals and terrorist may get access to it. Another major apprehensions of west about Pakistan nuclear arsenals is that it can transfer the technology or sensitive information to other countries as they have seen in the case of Dr. Qadir Khan's case of proliferation of nuclear network. (Cohan, 2011)

So, Pakistan should take pre-cautionary measures to deal all these threats particularly theft by terrorist elements in the region or transfer of technology incident like Dr. Qadir khan case but should maintain the balance of deter in the region particular towards Indian hegemonic behavior.

(B) Internal Security Patterns

Pakistan national security patterns have totally changed after the devastating event of 9/11. Now the concentration of Pakistan security is "from within" rather than external factors. These internal security concerns are instigated by resurgence of radicalism, ethnic diversification, socio-economic difficulties, and political instability.

- **Sluggish economy**

Economic inconsistency is a critical threat to national security of Pakistan at internal level. Lack of proper economic development has caused the emergence of social problems just like ethno-sectarian elements which weaken Pakistan internally. Economic security means a country has strong economy having no external threat. For economic security it is essentially required a sound per-capita income, strong Gross National Products (GNP) and Self-sufficiency. In Pakistan, all the governments after independence are fail to create socio-economic equality in the society which threatens the internal security patterns of the country. (Singh & Kukreja, 2005)

Because of no proper economic planning particularly after 1960s, corruption and poor governance, Pakistan is under gradually increasing heavy debt and absolutely is relying upon foreign aid for financing the investment.

Pakistan is well equipped with natural resources but these are not properly exploited and affect the economic developmental process of the country. At International level, Pakistan industry is not warmly welcomed owing to its poor quality and high cost. Pakistan's investment climate is not suitable for the international investors over the previous years due to three prominent factors as social and legal security, law and order situation, and strong policy formulation.

Weak governance along with a poor law and order situation prevents the investors particularly foreigners and disturb the flow of economic development. The law and order conditions in Karachi, parts of Baluchistan, and FATA is still not at satisfactory level.

Huge level of corruption, high inflation, and repetition of subside; debt financing and preservation of huge standing armies are further contributing rudiments. Unfortunately, domestic economic policies and institutions are under the controlled policies of International Monetary Fund (IMF) and World Bank (WB), consequently Pakistan has no proper control over its domestic financial policies. (Yaqub, 2011)

Economic depressions with rapid increase in population emanates a lot of problems like social injustice, increasing unemployment, poverty, poor health and illiteracy in Pakistan. A nation cannot be said as developed unless high education standards and well-equipped healthcare facilities but unfortunately, these both factors are being bitterly neglected in Pakistan as just less than 2 percent Gross Domestic Product (GDP) is being spent on this important factor. (Haider, 2008)

Pakistan is basically an agrarian state and agriculture is the backbone of Pakistan's economy and source of strengthening the country's food security. But, the water scarcity has damaged the concept of agrarian country for Pakistan. It is strongly analyzed that Kalabagh Dam is technically much sound and economically much profitable for the country's downtrodden economy but it has been made politically controversial issue among the provinces. (Lodhi, 2009)

Another major threat of water scarcity to Pakistan's economy is from Indian production of water reservoirs more than their limits on Pakistani Side Rivers. Another drawback to Pakistan economy is the large share of budget spending on its defense but as analyzed by General (R) Talat Masood, a prominent analyst on security policy that spending on the Eastern (Indian) front has remained almost same, but the budget enhancement is directly related to new counterinsurgency requirements.(Haider, 2010)

Energy crisis has lit the fire to economic depressions. Although Pakistan has huge hydro-power potentials and proven coal deposits but it has been facing the problem of energy shortages. But now government is trying hard to enhance power productions. A number of projects have been initiated to enhance power generation like 969MW project of Neelum-Jhelum, 1410 MW Tarbela 4 extension, 7100 MW Bunji, 430 MW Dasu, 740 MW Munda Dam, and 4500 MW Diamer Dam.7 In addition, the inauguration of Iran-Pakistan pipeline which is expected to progress the situation as the pipeline will be completed and the gas will be started to come to Pakistan. (Cohan, 2011)

- **Weak governance and failures of state institutions**

Since from the its independence, Pakistan is facing numerous challenges relating to insecurity in provinces such as sectarian violence, political instability, weak law and order structure, not easy access to judicial system, tribal system, feudal system and so on.

It is said that democracy is absolutely based on two essential doctrines i.e. "participation and accountability" and there is no governance if there will be no democracy. Unfortunately, if we peep into Pakistan's history, it is depressed from both principles as the poor enactment of accountability as well as minimal participation of general masses in governmental structure. (The Nation, December, 2010)

It is also stated that "It is not the gun but it is the man behind the gun that matter". Same case is with Pakistan political system in which every stake holder as common man, governmental officials and opposition is not sincere according to their mandate. It is the real base for letdown of Pakistani political system. (Ibid)

Justice for common people of Pakistan is still indispensable. Judiciary is one of the three pillars of the state. But in Pakistan political structure, there is an image of appointing or hiring judges rather than lawyers is necessary even to get justice. Furthermore, due to socio-political differences in the society, justice even not easy to attain. In reaction, the trust and faith of the common citizen has broken upon the judicial institutions3. It is crystal clear fact that although Pakistan has a written constitution but has been never capable of acquiring an autonomous, neutral and courageous judicial structure to gadget the basic rights of the common man as preserved in its 1973 constitution. (Ibid)

Another troublesome hurdle in the way of good governance in Pakistan is poverty which is owing to the mounting economic inequalities in the country. The Pakistan economic Survey 2009 - 2010 shows that the gap between the rich and the poor has been widened in the country in the periods of 2005 and 2008. (Mughal, 2014)

Shortage of transparency is another factor involves in the vulnerability of governance in Pakistan. It is more likely to lead it to vague and non-participatory decision making, inept project implementation and unrestrained financial corruption of the institutions. It also leads to control of bureaucracy upon politics and non-functioning of democratic culture in the country. The issue of corruption in Pakistan is considered a sole factor of mal-administration in the country because it occurs at the high level of management in the state.

The civil services of the country have built up highly centralized administrative setup which is neither appropriate nor very effective. Civil services in Pakistan are highly polluted from Political influence that has depressed the competence of state institutions positively3. There is a considerable need to boost up Pakistan's administrative efficiency and effectiveness.

The education system of the country is always been ignored in Pakistan and state machinery concentrate only upon political and economic sectors. It is fact that an educated population has potential to maintain political stability and economic development in any state. (Dawn, July 23, 2009)

But, unluckily this connotation is not so impressive for policy formulator in Pakistan where still the local citizens are agonized frequently and deprived of beneficial education system. A blend of factors like ill-responsible behavior of government's personnel, futile and worthless educational policies, lack of proper research work, unqualified teaching staff, and wide gap between students' and teachers' relationships, poor quality text books involves in this flawed educational system.

Another dent to good governance in Pakistan, the corrupted image of police department in the society which is an open truth. Every activity in police station is performed just to charge the money and due to it every illegal act becomes legal. (Mughal, 2014)

- **Terrorism, Extremism and Sectarianism**

Terrorism and extremism is considered a non-traditional security threat in Pakistan particularly after 9/11incident. Pakistan as being a frontline state in this war has to face more adverse security repercussions. From that devastating incident, the security patterns of Pakistan is still deteriorating day by day.

Terrorism is deep rooted in Pakistani society due to blend of factors like poor governance, deprivation of rural areas, inaccessible justice and weaponization. In addition, geo-strategic position of country has lit the fire to terrorist threats. Historically, the Iranian revolution and its counterrevolutionary powers, Afghan Jihad slogans and the Taliban Regime have adversely impacted Pakistan society. In 1980s, General Zia's policies of Islamization gave new directions to Pakistan society. (Afzal, et al., 2012)

During that policies many religious institutions (Madrassas) were established and used as epicenters for Afghan Jihad. Hence, Jihadi culture penetrated in our society and succeeded to create a favorable environment from political and religious parties. Then the episode of 9/11 came and Pakistan suffered a lot as being the partner of US in war against terrorism.

9/11 was only one terrorist activity inside the US territory but Pakistan has been facing a lot of such incidents in its society resulting in the deterioration of law and order and losses of general public. Even not any part of country is safe from terrorist or suicide attacks. Foreign and local Investors are reluctant to invest due to poor law and order situation and instability in the country. In addition, feeling such insecure and the poor law and order situation, the Pakistani intelligentsia is migrating for safe future to other countries like Canada, Australia, America, England, United Arab Emirate and South Africa and it is estimated roughly six million Pakistani intellectuals has shifted in the last three decades. (Askari, 2008)

Sectarianism is also another major internal threat to Pakistan's security paradigms. It has gradually disturbed the basis of Pakistan and has created the sentiments of violence, hatred and disorder in the society. Historically from preceding three decades sectarian clashes have augmented the suicide violence, bomb explosions, murders, and terrorist attacks.

After the Soviet incursion in Afghanistan, the clumsy issue of sectarianism has arisen and divided the society. Saudi Arabia was financing the Sunnis whereas the Shia sect was aided by Iran. A so-called ideological difference has emerged and has given the birth to sectarianism in the society and many lives have been lost due to this intrigue issue. The South Asia Terrorism Portal 2009 report from to

2002-08 tells us that there are 642 incidents of sectarianism in which 1518 people had lost their lives whereas the 2817 were injured. (Afzal, et al., 2012)

These religious extremists get education from their ideology oriented Madrassa and these Madrassas exploit them against each other and also provide arms for militant activities. The immediate threat of these violent activities affects the economy as the investor are reluctant to invest in Pakistan. The state is incapable to secure its citizen which automatically resulted in the "Privatization of Security" where "Small Arms" and "light weapon" have become foremost weapon of ferocity by both state and non-state actors which endangered the lives of innocent people. (Cheema, 2006)

Currently, Pakistan has become safe haven for manifold militant Islamist insurgencies. It had become the main cause of ferocity and political instability in the areas of FATA, Khyber Pakhtunkhwa (KPK) and in Baluchistan. Pathetic governance and over-emphasis upon military solutions have lit the fire and has boosted up the extremist feelings. Such kind of dealing the situations has become cause of connectivity between Al-Qaeda and Pakistani Taliban which has enhanced the suicide attacks and bomb blast in the society.

Tribal people of Pakistan particularly in FATA are suffering from hand to mouth for basic necessities of life and can easily be exploited by the militants and this sense of deprivation plays a vital role in the growing militancy in the Pakistan society. (Ali 2011, 76)

FATA is strategically more important for Pakistan because security of this region is necessarily considered the security of the whole country as it is considered as the gateway to Pakistan's territorial reliability and governance. (Hilali 2010, 67)

The critical flow of suicide terrorism was initially started in 2004 when an armed conflict was at its epics in Waziristan. The initiation of military operations on the Pak-Afghan border to crush these terrorist elements has resulted the enhancement of suicide attacks by these insurgents and even no place in the country in being safe these attacks. (Chandran & Chari, 2011)

After 9/11 incident, many Uzbeks, Tajiks, Yughiers, and Chechens terrorists penetrated into Pakistan and joined hands with Pakistani

Taliban and started terrorist activities to destabilize Pakistan. They are the sole responsible of drug trafficking, weaponization of society, kidnapping of people for money and also striking security forces deployed at the tribal areas. (Yousafzai, 2011)

The war on terrorism has suppressed a nuclear country and has brought it to its country to its knees. Although, Pakistan has sacrificed a lot in this global war but it could not get the recognition of its sacrifices but has always been bitterly criticized by Americans. (Rehman, 2011)

- ## National Identity Crisis and Provincialism

Pakistan is basically an ideology oriented country and is appeared at world globe on the basis of 'two-nation theory'. There are three basic foundations of Pakistan's identity as a nation such as the ongoing deliberations over the meaning of Pakistan and meaning of Pakistani, the special effort of collaboration of this identity with Islam, and communal, provincial challenges to the Pakistan as a nation. The rumors of a new disintegration of Pakistan on the basis of ethnic conflict are near to reality in near future. (Ayres, 2009)

Pakistan has multi- ethnic and multi-linguistic society and is based on many groups. Some of which have characteristics of nation although they have their own linguistic, culture and even identity but Some surveys show that these Pakistanis consider themselves as Pakistanis first and other identities like Punjabi, Baloch, Sindhi or Mohajir as second. (Cohan, 2011)

There is a lot of diversification among these ethnic groups as the Baloch are basically a tribal society, Sindhi are mostly rural, and the Mohajir are pre-dominantly urban. Sindh, also the most industrious province of Pakistan was divided between urban wealthy part and rural deprived segment of society. According to 1981 consensus, Sindh province had the 19.3 million population in which 10.6 million (55.7%) are native Sindhi, 4.6 million (24.1%) Urdu speaking Mohajir, about 2 million people are (10.6%) Punjabis, 1.1 million (6%) Balochi and so on. *(Ahmer, 1998)*

So, most migrants are being settled in the urban Sindh, consequently, the ingenious Sindhi have become numerically minority in Sindh's two largest cities, Karachi and Hyderabad. This situation started creating tensions in the Province. The Mohajir were in favor of Urdu as their official provincial language the Sindhi natives were demanding for Sindhi language. So, a tug of war has been started between two ethnic groups upon each issue like jobs quota etc. (Kennedy, 1991)

The issue of Baluchistan province is the complaints for the denial of political and economic autonomy. They are demanding for re-adjustment of power and refusing to accept of dominancy of any other non-Baloch community as Punjabi, Pashtuns or Mohajir. *(Khan, 2002)*

All these groups are using ethnic element as tool and exploiting their people for Punjab dominancy over them and demanding for their rights. Actually they are being used by some external agencies to who are against Pakistan identity and working to resurgence of 1971 story. Some of these ethnic groups are demanding for the proclamation of new provinces on the basis of these ethnic identities which will never be suitable for Pakistan national integration.

Historically, in the story of separation of East Pakistan, the West Pakistani elites underestimated their capability to curb the Bengali uprising and did not imagine that India could militarily to support it. But, today if we calculate ethnic and provincial slogans, we see Pashtun nationalism though disturbing but possibly cannot be a strategic threat to the state. Baloch movements historically were threat for country integration especially in the side of the military, but have been lessened with a combination of money and military actions. The demanding of a Saraiki province has not yet much impetus. And the demand of Hazara province have much little political influence.

All these movements of ethno-linguistic and provincialism are not much disturbing issues than organization of new Pakistani Taliban group comprised upon the blend of religious, territorial, linguistic and cultural commonality in the Pakistan especially in KPK.

The Pakistani Taliban are exploiting the deprived Pashtuns and portraying themselves as religious rather than ethnic terms because they think that religious mobilization is more impressive than ethnic mobilization. (Cohan, 2011)

- ## Demographic issues

Pakistan is one of those countries which are continuously booming with population day by day. Pakistan is becoming one of countries like Nepal, Yemen, Afghanistan and Congo where population increases to its double after one generation or almost after every 30 to 35 years.

British Council Study reported in 2009 that half of Pakistan's population is less than twenty three years age. Birth rates has become very high especially in rural areas. The population of Pakistan has grown up thrice time in less than fifty years and is expectedly to increase to approximately 85 million in the subsequent twenty years. (Ibid)

The rapidly growth in population will automatically need extra food, extra energy, and for males, more jobs. Especially currently when Pakistan is becoming more urbanized. As increase urbanization of Pakistan society is also a source of ethno-cultural rivalries in the society. As in Karachi, there is a high level of tension between Sindhi and non-Sindhi people and recently between influx of migrant Pashtuns from KPK and Sindhi. Intolerable safe-houses of extremism have been successfully maintained throughout Pakistan, particularly in the big cities as being mixing with ethnic enclaves and it could not possible for security forces to suspect them successfully.

It is also the question of the about the issue of alleged demographic dividend as whether the population growth is better for the country or a threat to its national security. Some people consider high population growth as catastrophic but some also argue that the population growth is not a threat but is a challenge and if it will be tackled properly, it will surely be fruitful for the country. (Cohan, 2011)

- ## Natural Calamities

Environmental calamities like water shortage, soil depletion, and natural tragedies can exaggerate rivalries or create pressure within a country and increasingly enhance to national security questions. (Center for Naval Analysis n.d) The climate change as considered a big threat to international stability and is also main factor which affected the Pakistan's national security. Climate change has increased the problem of global warming and bitterly changed weather patterns. Climate change causes possible threat in predictable future to Pakistan's population living alongside of the coastal line.

Pakistan is facing various types of natural calamities with some heavy aftereffects, particularly earthquakes and floods in the country. Pakistan has faced most critical earthquakes which resulted in substantial mortalities and destruction to infrastructure. (Pakistan's Current Monsoon leads to an International Emergency Appeal for Assistance, n.d)

So, such kind of natural disaster has created issues like food security, health issues and diseases, water and sanitation and infrastructure. Pakistan is facing an evil threat of flooding in monsoon season which mostly destroy the infrastructure as well as potentials of the national economy. Thousands of people are forced to leave their houses, crops and other necessities of life, in which many are killed due this natural calamity. A study regarding South Asia about the damages of flooding during 1990 to 2008 are as natural disasters killed 60,000 and affected 750 million people with $45 billion in economic losses in South Asia.

Natural calamities are mostly resulted in a large level of displaced populations and damage infrastructure in countries such as in Pakistan. But, if the government response is not so cooperative, than the negative element if the society attract these deprived and displace population and offer them opportunities of life but in return these deprived persons have no choice to follow their irrational commands. Same situation is in Pakistan particularly flooded areas of FATA and KPK. (Yusafzai, n.d.)

- **Media Role**

Media is also another imperative national security apparatus that influences to everyone in Pakistani society. The electronic broadcastings are being used at national and international level in order to stimulate public opinion. Because of weak political and governmental institutions in Pakistan, the media power is considered as a powerful device to control public agenda to reshape political settings according to one's own wishes. Although, in Pakistan media has been a sound instrument to criticize the regime's unlawful policies but sometimes it has also played a negative role against country's national interests. Shafqat Mehmood a famous analyst opines that media concentrate profoundly only on government and political issues, and does not focus upon the issues of community welfare. Regrettably. A planned and internationally command performance is being pursued by media on many fronts to degrade Pakistan's status globally. (Mehmood, 2010)

Few years ago, Pakistan had used state-controlled media to broadcast news to public according to its suitability. Prior to Benazir's reforms, the Pakistani media was forcefully state-controlled and all the media outlets like television and radio were broadcasting the command performance. Presently, media is almost free to on air everything suited to its agenda. But, Pakistani media particularly cable network does not focus on quality analysis and due to policy of liberalization of press, it undermines signs of the national and social welfare. These media network is absolute self-censored and mostly influenced by its own commentators who are doing most of the times directed performances. (Sabir, 2012)

Conclusively, media has an affective role in Pakistan society but it has necessarily to take some bold decisions regarding playing of positive role in the issues related to country's security patterns.

Policy recommendations and Conclusion

The comprehensive study of Pakistan national security patterns depicts a bleak picture of insecure state and society. The most cracking threat to its national security emanates from "within". Pakistan can't ignore the mounting challenges related to its sluggish economy, dispersed society, national character and morale, the issues distressing its national security.

Pakistan necessarily should take progressive measures of improvements to its "strategic culture" by adopting an effective approach to maintain its sovereignty and reliability and stimulate security. Stimulation of an institutional approach is urgently needed to draw the country out of the existing challenges. The institutional process will be helpful to install the proficient and honest people at the helm of powers to guide the country on the way of peace and prosperity.

Thus, the government should revise and modify the conception of National Security Council (NSC) as a genuinely operational body which might take all stake holders in the society on board to solve all the security issues in the state.

The foreign policy of Pakistan should be centered on common coordination on the standards of "friendship with all enmity with none". On the issue of nuclearization of the region, Pakistan should remain in favor of universal nonproliferation efforts but on equal basis and clarify the whole world that its nuclear capability is just to create Deterrence in order to avoid regional imbalance which is a severe threat to its security apparatus. Consequently, Pakistan may sign all the non-proliferation treaties including CTBT if all the contestant stakeholders particularly India, are ready to sign it.

The resolution of Afghanistan crisis is enormously essential for regional stability. In order to maintain stability in the region, Pakistan should adopt the negotiation approach to solve all the disputes that can lead the region towards economic development and peaceful coexistence.

It is also proposed by many security analysts that if Pakistan is to concentrate on combating the extremists "within", who are the major national security threat, it is needed to lessen the tensions with India in order to avoid a two-front military deployments.

Pakistan's geo-strategic position offers enormous prospects. Peaceful security surroundings at its eastern and western boundaries is an enormous arrangement for the emancipation of economic opportunities. The Foreign Office if the country should find out the options to acquire full affiliation of Shanghai Cooperation Organization (SCO) and South East Asian Nations (ASEAN), the sole hub of economic activities in order to boost up its economic settings.

Pakistan can still positively enhance the mutual cooperation among the Muslim countries by boosting up the role of Organization of Islamic Conference (OIC), a significant forum to bond the Muslim Ummah. However, Pakistan must undoubtedly declare its sociable countries to avoid meddling into its internal matters in the shape of direct financing to various contentious groups in Pakistan.

Furthermore, Pakistani Foreign Office must voice at each forum including the Unite Nations General Assembly for losses that Pakistan is facing while playing the role of frontline state in the global war against terrorism. International community must acknowledge Pakistan' efforts and damages and come forward to support the country economically then Pakistan can affectively play a role of bringing permanent peace and stability in the region positively.

Pakistan's economic and societal development is closely interconnected with democratic system of the country. If the democracy is workable properly, it may bond of both person and purpose and galvanize people. Therefore, at the first step, all political parties must carry out the democratic values within their parties as to democratically legitimize itself to strengthen the democratic culture. Then, the Pakistan's leadership should sit collectively to progress a unified and logical strategy to stop the spread of anarchy everywhere in the country.

Economically, Pakistan cannot afford the culture of provision of subsidies and should not simply try to qualify for another IMF installment because more debts will also lead to lessening foreign exchange assets and resultantly, put burden on the exchange rates and create inflation in the country. The possible solution is to enhance resource utilization domestically and specific taxation measures. But, precautionary measures must be taken to stop tax leakages through corruption.

Terrorist threat to Pakistan national security is domestic as well as trans-national phenomenon. But, an impressive approach is necessarily needed to address the causes of this violent threat and to explore most effective way to curb these causes positively. But, the most important thing is that we cannot expect that problems that took a lot of time to penetrate in the society, get resolve immediately, it may proceed for weeks, months, or even years.

The radicalization of society is also another serious threat to Pakistan's security apparatus. But, it can be washed out as government should encourage common people to step forward to help the government. It is the best way to eliminate the radicalization tendencies from the society.

Corruption is also a negative element in weakening the system of governance in Pakistan, to curb this problematic issue, it is strongly needed to focus on structuring of institutions and systems such as accountability mechanism to check the responsible. Furthermore, the country's anti-corruption policy that was announced during 2002 but remained in letters only, must be prompted in letters and spirits.

Devoid of considering of the health and education sectors of the society, one cannot prepare a healthy and visionary nation that is compulsory for country' prestige and prosperity. So, if Pakistan wishes to maintain its nation on the right way, it will have to review its educational policy. Discriminatory level of education will lead to the creation of sentiments of hatred and intolerance in the society. Additionally, Madras and public school sectors must be given more attention. Nonetheless, to boosting up the level of these two sectors of society, a considerable budget must be allocated on emergency basis.

Furthermore, the increasing of urbanization of the Pakistan population particularly in mega cities, there is no immediate formula for it however, short term measures can be taken to lessen the issue as to generate inducements for the population to stop them from migrating to mega cities by facilitating them equitable necessities like education, health and employment opportunities at their doorstep.

Addressing the issue of energy crisis, it is necessary to create a sense of realization in the society that if there is no energy, then there will be no national security. The concerned departments must do the needful to stop the line losses and energy theft. At the governmental level, the leadership should make necessary arrangements to address the issue on emergency basis and do the needful to solve the issue on all levels as short term as well as long term accordingly. Likewise, the nation can also participate in curbing the problems at domestic level by minimal and compulsory usage of energy.

Most intensely, there is a strong need for maximum usage of accessible water by producing reservoirs. A comprehensive and obvious approach is necessarily needed from all level of society to address the problem of water scarcity.

For natural calamities, National Disaster Management Authority (NDMA) should take pre-cautionary measures to face the expected potential threats from the nature. To face and handle the issue properly, a coordinated approach is necessarily needed by NDMA with all other forces like the NGOs. National Disaster Management Commission's (NDMC), representatives at various tiers to utilize the maximum available resources positively. Government must allocate mandatory financial resources for the disasters. A well-established "disaster management system" must be created because sometimes, the availability of the armed forces is not possible due to its installation in other national exigencies. And lastly, media is needed strong overhauling because it has the capacity to do a lot to renovate Pakistan's image at international level. It is suggested that media should not hide the bad but also describe the good to balance it. Pakistani media must boost up its capability to reach the level to differentiate between constructive and negative criticism. There must be code of ethics for media which should be maintain by Press Council of Pakistan positively.

Conclusion

Pakistan's security pattern are multidimensional and has remained in dynamic positions at both levels internal as well as external to deal massive challenges, Pakistan should adopt consistent grand strategy. Historically, country has been involved to adopt policies at one form to other form just to deal with the situation at the spot and no national policy has been opted for long terms to address the security patterns positively. So, the result has always been in the dismemberment of half country. The antagonistic forces are still working for success in their unfinished agenda. Therefore, Pakistan's national security apparatus cannot afford any deficit in its security paradigm. Consequently, a long term, well-equipped and coherent grand strategy is needed to boost up country's national defense internally as well as externally.

Pakistan's leadership has to work charismatically to serve the nation on the basis of two options: either to serve the nation honorably or leave the helm of power honorably as competent and sincere people to move ahead to steer this country out of the recent threats which contains all indications of a "failed state". In the formulation of coherent policy for strengthening the country's national security, it must be followed that it should address the issues particularly at domestic level and also keep in mind the external influences as the global actors influence according to their vital interests. And lastly, the grand strategy must have the options of modifications as with the passage of time and emergence of contributing factors to these threat should be addressed positively.

References

1. Afzal, S., Iqbal, H. & Inayat, M. (November – December 2012). Sectarianism and Its Implications for Pakistan Security: Policy Recommendations Using Exploratory Study. *IOSR Journal of Humanities and Social Science (JHSS) Vol.4 (4).*

2. Ahmar, M. (1998). *Internal and External Dynamics of South Asian Security.* Karachi: Feroz sons Pvt.

3. Ali, A. (spring & winter 2011). Internal Instability in Pakistan-Ideological and socioeconomic perspectives. *Strategic Studies Vol. 31(1&2).*

4. Askari, J., S. (October 9, 2008). Economic Woes Cause Brain Drain. *The Nation (Islamabad).*

5. Ayres, A. (2009). *Speaking Like a State: Language and Nationalism in Pakistan*. Cambridge U.K.: Cambridge University Press.

6. Baylis, J. & Smith, S. (Eds.). (1997). International Security in the Post-Cold War Era, in *the Globalization of World Politics.* Oxford: Oxford University Press.

7. *BBC.* (April 18, 2010). Global Views of United States Improve While Other Countries Decline. Retrieved from http://news.bbc.co.uk/2/shared/bsp/hi/pdfs/160410b bcwspoll.pdf

8. Brigadier Khan K., M. (2011). A Stable Pakistan: Proposed Model of National Security. *Margalla Papers Vol. XV (II).*

9. *Center for Naval Analysis.* National Security and the Threat of Climate Change. Retrieved from http://securityand climate.cna.org/.

10. Chandran, D., Suba, and Chari, P., R. (2011). *Armed Conflict in South Asia 2010: Growing Left Wing Extremism and Religious Violence.* New Delhi: Routledge.

11. Cheema, I., P. (2006). Political Violence and Terrorism in South Asia. *Islamabad Policy Research Institute (IPRI) Publication.*

12. Cheema, I., P. (July-September 2006). Pakistan-Afghanistan Relations. *South Asian Journal* Vol. 13.

13. Cohen, P., S. (2011). *The Future of Pakistan.* Washington, DC: Brookings Institution Press.

14. Cohen, P., S. (2004). *The Idea of Pakistan.* Washington, DC: Brookings Institution Press.

15. *DAWN.* (July 13, 2009). Ex-generals, MPs want Military out of politics.

16. *Dawn.* (July 17, 2007). Vulnerability of Al Qaeda leaders has risen: Report. Retrieved from

17. *Dawn.* February 27, 2013.

18. Dr. Haq, U., N. (Ed.). (April 18, 2013). Pakistan-India Peace Process 2008-2009. Islamabad: *Institute of Policy Research.* Retrieved from http://ipripak.org/factfiles/ff117.pdf

19. Dr. Lodhi, M. (July, 14, 2009). Pakistan's leaderless moment. Islamabad: *The News International.*

20. Haider, M. June 11, 2008). Govt to endorse poverty figures of Shaukat regime. Islamabad: *The News International.*

21. Haider, Z. (June 5, 2010). Militancy-Hit Pakistan Ups Defense Spending by 17 Percent. *Reuters.* Retrieved from http://www.reuters.com/article/idUSTRE6541UF2010 0605.
22. Hilali, A., Z. (2010). FATA: The Strategic Depth of Pakistan. *Margalla Papers.*
23. http://www.dawn.com/news/256761/vulnerability-of-al-qaeda-leaders-has-risen-report
24. http://www.ipsnews.net/2005/11/pakistan-militants-gain-legitimacy-through-quake-relief/
25. Hussain, Z. (2007). *The Struggle with Islam*. Lahore: Vanguard.
26. Jaspal, N., Z. (Winter 2010). Threat of Extremism &Terrorist Syndicate beyond FATA. *Journal of Political Studies, Vol. 17 (2).*
27. Kennedy H., C. (October 1991). The Politics of Ethnicity in Sindh. *Asian Survey Vol. XXXI (10).*
28. Khan M., M. (Feb. 2014). National Security's Dimension & Challenges: The Way Forward. *PIPS Research Digest- Vol.1 (2).*
29. Khan, A. (March/April 2002). Pakistan's Sindhi Ethnic Nationalism: Migration and Marginalization, and the Threat of 'Indianization'. *Asian Survey Vol. XLII (2).*
30. Khan, I. (June16, 2014). All-out military operation launched in North Waziristan. *Dawn*. Retrieved from http://www.dawn.com/news/1112949
31. Kukreja, V. &Singh, M., P. (2005). *Pakistan: Democracy, Development and Security Issues*. New Delhi: SAGE Publications.
32. Kux, D. (2001). *The United States and Pakistan, 1947-2000: Disenchanted Allies.* Washington, DC: Woodrow Wilson Center Press.
33. Lodhi, M. (2012). *Pakistan: Beyond the Crisis State*. Washington, DC: Oxford University Press.
34. Mahmood, S. (December 24, 2010). Media and the nation. Islamabad: *The News International.*
35. Miller, E., S. (Summer 2001). International Security at Twenty-five: From One World to Another. *International Security, Vol. 26 (1).*
36. Mughal, R., M. (February, 2014). GOOD GOVERNANCE IN PAKISTAN PROBLEMS AND PROPOSED SOLUTION. *International Journal of Modern Business Issues of Global Market (IJMBIGM) Vol. 2 (2).*
37. Pakistan's Current Monsoon leads to an International Emergency Appeal for Assistance. (n.d.).Retrieved from *http://www.examiner.com. http://www.examiner.com/article/pakistan-s-current-monsoon-leads-to-an-international-emergency-appeal-for-assistance*
38. Prof. Dr. Musarrat, R., Afzal, R. & Azhar S., M. (2013). National Security and Good Governance: Dynamics and Challenges. *Journal of Public Administration and Governance, Vol. 3 (1).*
39. Rehman, U., A. (October, 2011). Pak-US Strategic Relations under War on Terror A Decade of Security Challenges for Pakistan. *Monthly Current Affair Digest.*
40. Sabir, M. (2012). Analysis of mass media: Influences on Socio-political perceptions and stability in Pakistan. *Journal of Political Studies, Vol. 19(1).*
41. Shah, S. (June 17, 2014). Pakistan troops killed in assault on Taliban. *The Wall*

Street Journal.

42. Shaukat, S. (August 3, 2009). India initiates deadly arms race *The Frontier Post.* Retrieved from http://www.thefrontierpost.com/News.aspx?ncat=ar&nid=476

43. Singh, M., P. and Kukreja, V. (Eds). (2005). *Pakistan Democracy, Development and Security Issues.* New Delhi: Paramount Publishing Enterprise.

44. Smith, S. (1999). The increasing insecurity of security studies: Conceptualizing security in the last twenty years. *Contemporary Security Politics Vol. 20 (3).*

45. *South Asian Terrorism Portal 2009 Report.* (July 31, 2012). Retrieved from http://www.satp.org/satporgtp/countries/pakistan/index.htm

46. Syed, M. (Summer 2014). Paradigm shift in Security: Adopting Human Security in Pakistan. *IPRI Journal Vol. XIV (2).*

47. *THE NATION.* (December 10, 2009). Combating Corruption.

48. *The News (Lahore).* (May 31, 2009).

49. Tzifakis, N. (2011). Problematizing Human Security: A General/Contextual Conceptual Approach. *Southeast European and Black Sea Studies Vol. 11(4).*

50. *U.S. Agency for International Development.* South Asia Earthquake Fact Sheet #44.Retrieved from http://www.usaid.gov/our_work/humanitarian_assistance/disaster_assistance/countries/south_asia/fy2006/southasia_eq_fs44_08-25-2006.pdf.

51. *UNDP. Human Development Report.* (1994). New Dimensions of Security. Retrieved from http://hdr.undp.org/en/media/hdr_1994_en_chap2.pdf

52. Yusafzai, I. (n.d.). Militants Gain legitimacy Through Quake Relief. Retrieved from *http://ipsnews.net.*

53. Yusafzai, I., H. (June 2011). *The US Factor in Pak-Afghan Relation post 9/11.* MB & Co. KG Germany: Lambert Academic Publishing (LAP).

54. Zaidi, M. (February 22, 2011). Kicking the habit. Islamabad: *The News International.* Retrieved from http://thenews.jang.com.pk/TodaysPrintDetail.aspx?ID=32466&Cat=9

55. Zaidi, M. (2012). *Insight on Insecurity in Pakistan.* New York: Narratives,

CHAPTER NO. 3

PAKISTAN'S EXTERNAL SECURITY CHALLENGES

Introduction

"Naturally, the common people don't want war, but after all, it is the leaders of a country who determine a policy, and it is always a simple matter to drag people along whether it is a democracy, or a fascist dictatorship, or a parliament, or a communist dictatorship. Voice or no voice the people can always be bought to the bidding of the leaders. This is easy. All you have to do is tell them they are being attacked, and denounce the pacifists for lack of patriotism and exposing the country to danger. It works the same in every country. " (Hermann Goering, Hitler's Reich- Marshall at the Nuremberg Trials after WWII)

The world woke up to learn about the new external security threats after 9/11.

History of the world took a sharp twirl after the US began the first war of the 21st century, something which would come to be known as the War on Terror. President Bush Jr. statement, "We will make no distinction between the terrorist who committed the crimes and those who harbored them." (9/11 Comission Report, 2004) turned the world turn upside down. The international security system following the three principle rules of statism, survival and self-help made United States take decision, which it justified as the need of the circumstance. From the time since the attack, Pakistan's external, regional and bilateral security challenge have evolved, throughout the transnational, traditional (global, regional and bilateral) and non-tradition confronts. These challenges have gravely jeopardized the preservation of national sovereignty and the territorial integrity of the state and, so, would be the focus of discussion in this chapter.

The International Environment - Pakistan's Existential Threat

The international system lacks to it the power of solving the problems of the world, which is the ground of strategic fluctuation and tentative transitions. Confronting to the external challenges are the imperative internal problems., (Lodhi, 2012) The two being inseparable will need to be handled with heed, simultaneously.

The nature of the security challenges which, in the past centuries, used to take into contemplation only military power, now has added to it transnational threats . These take account of population growth, distressed migration /flow of refugees, Transnational Organized Crime (TOC), transnational terrorism, scarcity of resources, proliferation of weapons, drug and human trafficking, and the emergence of non-state actors and asymmetric warfare, among others.

Existential threats shape the country's traditional security posture, and Pakistan is no exception to this phenomenon. Pakistan's geographical location and its position on international alliances have forced it to make some unpopular choices.

Pakistan's Strategic Culture

When a state becomes exposed to internal violence, dismemberment, internal instability and external interventions, it means that there is a solemn divergence over the purpose of the state or its ideology. In this situation, violence and struggle for power erupts within the state among the rivals, which in turn invites foreign interventions. Insurgents and holding regimes may then use external aid to support their ethnic, religious, ideological or strategic needs. (Budani, 2001)

Political and Institutional Hurdles in National Security Policy Making of Pakistan

Professor James Petras (2011) in his article, "Puppets in Revolt: Afghanistan, Iraq, Pakistan and the United States" highlighting the external structure of the world order lays down some of the features, which help in understanding the nature of the established security policy of a particular state.

- Neo-colonial puppet installation and the creation of the New World Order

  The beginning of anti-colonial movements against the western imperialism ultimately became a challenge for the US. Thus, in order to keep hold on its economic and military bases, it maintained 'political collaborators that would fulfill their agendas' through institutional continuity. In the name of 'foreign aid', an enriched and native kleptocratic bourgeoisie was created. By providing military aid, mission exercise and scholarships abroad, a generation of military and civilian bureaucrats is being created to serve imperialist world-view . By establishing the ideology of free market (capitalism) through Latin America, Asia , Africa and Eastern Europe , the culture of pillage , concentration of wealth and privatization became prevalent. The neo-conservative ideology permeated with Israel's colonial mentality. The war in 1990 has made them believe that "empire building was the inevitable wave of the future. Only an appropriate political trigger was necessary to mobilize the financial and human resources to pursue the new military driven empire."

- Exploiting 9 /11 – Welcoming ' Shadow Puppets"

Known to be as "a word-wide military crusade against terrorism," through allocation of massive funds for 'media propaganda', the designs of new imperialism began in 2001 – invasion of Afghanistan by US-NATO mercenary armies and defeat of the Taliban. Invasion led towards regrouping of the anti-colonial forces, today the outcome of which is civil unrest in the country and its "domino effect" in the neighboring country, Pakistan. However , all these wars and occupations has led to US bankruptcy and since the strategy of invasion is not welcomed by people of the US , one sees a period of withdrawal of troops, welcoming of prolonged periods of civil wars, domestic revolutions and democratic social revolutions in countries like Afghanistan and Pakistan and setting up of shadow puppets in developing nations across the world.

Asif Ali Zardari, a puppet head of the state planted by a sham democracy, faced opposition from the agencies of intelligence and the military. With 85% of the people being hostile to the US, the military of the country become involved in military offences. The fight against the Islamist communities left about 4 million people as internal refugees. Zardari lost his veracity because of escalation of the war to destroy the safe haven of Taliban and their Pakistani fortified partners. Empowering the Special Forces of the US to function from the local bases of Pakistan and everyday US bombing against the civilians developed among the national the universal consensus of his covert and puppet loyalty.

"While puppet rulers provide a useful façade for external propaganda purposes, *their effectiveness diminishes to zero domestically, as their subservience before the imperial slaughter of non-combatants increases*. The initial imperial propaganda ploy portraying the puppets as "associate" or "power-sharers" loses credibility as it becomes transparent that the puppet rulers are impotent to rectify imperial abuses. This is especially the case with pervasive human rights violations and the destruction of the economy. Foreign aid is widely perceived as nurturing widespread extortion, corruption and incompetent." (Prof. Petras, 2011)

- Global Alliance – A dilemma; Interventions and Backlash

The existential threats encountered by Pakistan since forever has made it look forward toward the US for arms. US, in turn, has used 'Pakistan's geographical position' against Soviet Union's expansionist policies and Al-Qaeda/ Taliban forces in Afghanistan.
Today, the progressive China-Pakistan partnership has gained Pakistan support against its rival, India (Oldenburg, 2010) indicating the gradual shift and tilt of policy, towards the next predicted emerging power, China.
As has been observed, in the history of Pakistan, the process of democratic transition has been overshadowed by the role being played by international powers for their regional and global interests, i.e. U.S. support of military coups in Pakistan's history. (Cohen, 2004) US/NATO intervention produced backlash among the citizens in violation of territorial integrity and sovereignty inviting anti-American sentiments. Pro and anti-democratic forces have used these opinions and belief formulated, to play their role as either 'shadow puppets' or to achieve the agenda of a particular party. (Anderson, 2012)

- The Role of Military and Civilian Institutions in Politics

If there is no security, there is no state, there is no culture and there is no society. For national security to be guaranteed, a state needs to domestically 'organize power.' In case of Pakistan, until today, with feudal mentality still customary, and governance of authoritarianism still prevalent, civilian institutions deficient of a strong organizational structure are struggling to establish their role in the society. In contrast, since partition, the national security structure of Pakistan, (military and bureaucratic) has been chosen to serve the remainders of the colonial rule (Hafeez, 2012). Throughout the history of Pakistan, for survival and against country security threats, military coups have filled the vacuum. If Pakistan's civilian government and military continue in the same direction, the country should be seeing more of disharmony in civilian–military relations in coming years, resulting in policies which are either positive or negative, endangering sovereignty, national honor and territorial integrity of the state. The government of PPP under Asif Ali Zardari, for example, was unsuccessful in delivering prosperity, due to which there was political and social unrest in the country. During the regime of President Musharraf, under external pressure he had not only to 'separate his role as head of the military and nation, but had also to hold elections.' (Darby, 2013)

Security Concern's for Pakistan

¤ **Iran and Saudi Arabia-Proxy Warfare**

Introduction

Hassan Abbas in his article ," Shiism and Sectarian Conflict in Pakistan Identity Politics, Iranian Influence, and Tit-for-Tat Violence" has written extensively on this proxy warfare , of which has been studied and understood as written in the following paragraphs.

In countries like Saudi Arabia, Yemen, Bahrain and Pakistan, sectarian sadism between Sunni and Shia's seems to have intensified. The second biggest Shia community resides in Pakistan after Iran, in terms of numbers. Before the period of 1977, both the Shia and Sunni communities had cordial relations. However, the events that followed in the aftermath of the Afghan- Soviet war of 1980, the contentious 'Islamization' policies of Zia and the Islamic Revolution in Iran in 1979 provided for a sense of empowerment for religious freedom, making their loyalties to Pakistan a difficult undertaking. Having been persecuted in the hands of anti- Shia organizations, minority Shia groups acted in defense of their people. This engaged a tit-for-tat violence in which Pakistan became a country for a proxy Saudi-Iran war. After 9/11 attacks, these communities have been used against one another, further deteriorating the efforts to promote inter-communal harmony. The problem is thus two-fold:

1. Financial support for the two sects of Islam by Saudi Arabia and Iran
2. Presence of militancy and extremism has insecured the internal environment of the country.

According to the political scientist, Khan (2003):

"Shi'ism in Pakistan became more centralized, more clericalist, more Italianized, and more integrated with the international Shi'i community.

At the same time, Sunnis became, more Arabicized as a result of the mass migration of Pakistani labor to the Gulf States and the generous funding of Saudi Arabia to the Pakistani Sunni madrasahs and to the jihadi organizations." (p, 64)

A Brief History

First, in July 1977, when Zia ul –Haq came to power, his doubtful leadership and mala fide policy verdicts under the veil of Islamic reforms shattered the sectarian harmony . Then, in 1994, the increase of Taliban in Afghanistan, gave anti-Shai militants a safe heaven, which completely engulfed Pakistan in issues of sectarianism. (Abbas, 2010)

Situation after 9/11 attacks

After the 9/11 attacks, the militant groups of both Shia and Sunni communities were banned by President Musharraf, but since his end of tenure, knowing the lessons of failures of militancy, there are still fears lingering about the resurgence of the militant groups. In an interview conducted in Islamabad among the Pakistani Shia's about Musharraf's nine years in power , (1999-2008) in spite of other grave objections, was viewed as a positive period in terms of security, since there were crackdown's on anti –Shai organizations and counterterrorist operations against Al-Qaeda in the Pak–Afghan border areas.

The ongoing violence against the Shia community has been the operational direction under Al-Qaeda and Taliban to the Sunni sectarian groups, such as the SSP and LEJ (Raman, 2007). The distributions of propaganda materials spread in Pakistan have enclosed anti-Shia slogans. Establishments between the so-called Punjabi Taliban (with significant SSP and LEJ) participation along with the Pakistani Taliban stimulated these campaigns. (Abass, 2009)

The Response of Pakistani Shia Community

The Pakistani Shia community has maintained relations of goodwill towards Iran. However, Iran's labour to influence much of the rituals of their community in the region of South Asia has not been successful. Despite the fact that support campaigns by the ISO were ran, many Pakistani Shias ignored the fatwa's issued by Ayatollah Khomeini. (Pinault, 2003) Many educated Shias living in Pakistan have realized that Iran, for attainment of its own political goals has been using the Shia's residing in Pakistan. Furthermore, if Sunni turn onto militancy, Iran is of no support in protecting the Pakistani Shias (Interview, 2009).

Hasan (2007), having studied the dynamics of violent behavior, being present and lingering on in the Shia-Sunni hostility elaborated his view's in these words, "Sunnis and Shia, largely, do not hate each other." The matter of the fact is that "Pakistan is becoming a relocated battlefield for the Sunni-Shia violence of the Middle East," with the enlightened Sunnis realizing the sectarian game of Saudi Arabia in Pakistan, just like their Shia counterparts responded to such activities here sponsored by Iran.

Exploitation – Enemy States Entry the Society

In this battle of proxy wars, the role of the states of India and America has now been added, which has exploited the situation and taken advantage of the sectarian divide, devising strategies and propaganda to not only disintegrate Pakistan from within but to also damage interstate relations of Pakistan with Afghanistan and Iran among others.

General (retired) Mirza Aslam Beg describing in his article " Global Conspiracies Against Pakistan " has elaborated that, "Sarobi which is near Kabul is the main centre where an Indian General Officer in dominion of the Border Road Organization (BRO) has forward bases in Ghazni , Khowst , Gardez , Jalalabad , Asadabad , Wakhan and Faizabad .The instruction camps in Faizabad , having Muslim mullah , trainers and guards is disguised as a camp for the recusants of Chinese. Roads have been built from the areas of Sarobi, Asadabad and to Faizabad, by the BRO, from where targeting inside Pakistan is carried out. The network of Sarobi target's the province of NWFP (which is now Khyber Pakhtunkhaw).The region of Baluchistan is a being targeted from Kandahar. Balochistan dissidents are trained at Lashkargah. The along with the Balochistan Liberation Army (BLA) are specified the mission of targeting Chinese, who work at Gawadar, Sandak and Hab. From the anchorages, the Amercians and BLA undertake the jointly planned operations in Balochistan. From the outposts of Mand, operations inside Iran are also carried out. A warship in the Arabian Sea, intelligence bases in Muscat backup, assistance is being provided. What were provided as logistical bases of support for the American operation in Afghanistan, facilities at Jiwani and Kot Kalamat, are now being utilized for creating deteriorating situation in Balochistan and Iran." (The Nation, 2014)

A possible solution that can guide us in formulating our own policies is well framed by Riikonean (2007) in the paragraph below:

"The phenomenon of sectarianism goes beyond the different armies' as the competing militant groups call themselves and the efforts to curb sectarianism needs to reach further than controlling those groups. Equally significant is the need to look at the regional stakeholders when dealing with sectarianism, especially when sectarianism has acquired wider regional function in the area after the US invasion in Iraq."

¤ Indo-Israel Nexus : A Threat to Sovereignty and Survival of Pakistan

Introduction

The buildup of an alliance is a well-established tradition in order to maintain the balance of power in the anarchic world. One can easily historically observe how states can abuse international laws, commit crimes and be drawn in all kind of acts of atrocities. There are three powerful states allegedly acting to wipe out the existence of Pakistan today: the Indo-Israel-U.S. Triangle. According to Samuel Huntington,

"The most striking commonality is their perception of Islam as their common enemy. Their common target is to thrive on the illegal acquisition of wealth and the resources of the Muslim World. Both are working under the tutelage of the U.S. who is working on a similar agenda for the realization of its global imperialism."

A Brief History

The post- cold war developments lead to the beginning of a new Indo-Israel strategic partnership. The establishment of diplomatic relations between the two countries on January 29th, 1992, made India the strongest ally and collaborator of Israel (Halim, 2003).

Promotion of Indian Objectives in its partnership with Israel

Nasser (2007) wrote that Indo-Israel partnership had two objectives:

1. Overcoming the propaganda warfare on Kashmir by Pakistan
2. Finding for it, a trading partner, that could supply India with modern weapons; India, after the disintegration of the USSR, saw Israel as the best option

Post 9/11 Developments

- **Indian Objectives**
    1. Technological superiority against Pakistan.
    2. Israel could help India upgrade and modernize its armed forces.
    3. 'Akhand Bharat'.
    4. To achieve its national goals, in which Pakistan – India tensions continue to persist. It seeks to upgrade its capabilities and military technology.
    - Not at ease with the idea of a 'Nuclear Pakistan', that would seriously question the regional hegemony of India and its illegal occupation of Kashmir

Israel's Objectives

1. Military superiority against its neighboring Arab states and Iran.
2. Exporting arms – a security policy of Israel since 9/11 made it look for markets, to which could be exported three-fourths of its production and, thus, India became a market for the defense industry of Israel.
3. 'Greater Israel'.
4. Not at ease with the idea of a 'Nuclear Pakistan', for it fears that it would provide support for the people of Palestine, which is a security threat, for Israel. (Khan, 2011)

Similarity in Attitude and Occupation

Oppressive attitudes		
Hostility for their neighbor's	Arab-Israel Hostility	Indo-Pak Hostility
Forceful occupation of territories beyond their geographical boundaries	Israel-Occupation of West Bank and Gaza	India-Occupation of Kashmir
Nature of Foreign Policy	Israel-Investing in creation of an international environment, under which it can accomplish its goals	India-Investing in the creation of a regional environment, under which it can arrive at its goals

Source: Studied and compiled from the article, "Development in Indo-Israel Defence Relations Since 9/11: Pakistan's Security Concern and Policy Options," (2011) written by Zahid Ali Khan.

Changing International Environment

Realpolitik, required for envisioning the future strategy was realized by the Indian leadership. Having seen Israel as a state that would eventually help India toward establishing good ties with the US, Indo–Israel bilateral relations became stronger. (P, 2005)

War in Afghanistan, Iraq (Second Gulf War), Indian parliament attack in December 2001, events of 9/11, the crisis of Kargil in 1999 and the 2008 Mumbai terrorist attacks paved way for a greater Indo- Israel military co-operation. Support in logistics, surveillance, congregation of intelligence, joint exercises, adoption of Israel's guerilla strategy and its skills for fighting the Palestinian freedom movement has also been employed by India in the Occupied Kashmir. (Dudkevitch, 2003)

No Change of Indian Defense Policies

Knowing that Indo-Israel partnership has established for itself clear objectives, which have a huge commonality of interests involved, any change in Indian leadership will not affect the joint venture, unless their common objectives are achieved.

Here is how the Indian defense Minister, Pranab Mukherjee, had expressed in 2004, these ties, "There will be no change in the existing defense ties between India and Israel." (Indian Express, 2004)

In his address to the American Jewish Committee, the defense corporation has been outlined by the national security advisor of India, Brijesh Mishra, in these words:

"The democratic countries that are the prime targets of international terrorism should form an alliance and develop multilateral mechanisms to counter that menace. He identified India, the US and Israel as countries to form an alliance for this purpose. Such an alliance would have strong political will and moral authority to take bold decisions in the cases of terrorist provocation." (Sahgal, 2003)

Indian Defense Budget–2014: "There can be no compromise with defense"

India, from importing arms from Russia, has now moved toward the U.S. and is hurrying to update its military against a confident China and a rival Pakistan. India's recent upgrade of $ 3.5 billion to modernize its military hardware and armed forces, will primarily be more alarming for Pakistan than for China.

Here is what U.S. former ambassador to India, Robert Black Will said at a Council Meeting,

"Indians have no interest whatsoever in trying to contain China because they believe this could be a self-fulfilling prophesy, and their whole policy is to seek the best possible relationship with China," (The Washington Post, 2008)

Advantage for India – The Buy and Make Approach

Having India as the biggest market for exporting weapons, Israel easily diversifies its sources. Indian defense, modern capabilities, development and research would allow for the buildup of defense in the private sector of India. Hindustan Aeronautics Limited, aiming at assisting the private companies, chooses for foreign partners who can facilitate them build up the high technology weapons, including avionics and weapons system integration. (Khan, 2011)

Indo-Israel Nuclear Deal

Israel and India are not members of NPT. Neither do both the countries deny their nuclear co-operation dealings. Some of their dealings have included:

- Israel's Happy Missile – Gaining an attraction for the Indian Air Force (IAF), the missile can respond to the any electronic measures, which can be used to destroy the radars of the enemy.
- IAF areas of interest include, the Delilah II bombs, Pechora III , Popeye beyond - visual-range air-to-air missiles, surface-to-air missiles and crystal maze bombs.
- Sales of Arrow-II anti- ballistic missile defense system, by Israel to India. This is to make strong the air defense, which can counter the defense capabilities of China and Pakistan.
- Acquisition from Israel of "Green Pine," fire control radars, which being a important constituent of the Arrow system is used for the detection of missiles, arriving from a distance of 500km. (Pant , 2004)

Indo- Israel Navel Cooperation

Martin Sherman article, “From Conflict to Convergence: India and Israel forge solid relations”, published in 2003 details upon this cooperation, highlighted the following:

- Strategic alliance in the South Asian waters benefits both Israel and India.
- The dominant occurrence of the Indian Navy in the Indian Ocean allows Israel to institute its joint logistical infrastructure with India and test its missiles. As reported, in 2001, cruise missiles that were proficient to carry nuclear warheads were tested in the Indian Ocean, by the submarines of Israel.
- A deal to purchase, a $97 million electronic warfare system have also reached upon by both Israel and India.(Pant, 2004)
- From the fear of a long-range first strike and the protection of the Israeli installations based on land, use of a sea-strike second capability would provide for Israel an effective strategic deterrence. Indian Ocean , in this regards provides for a base where logistical infrastructure, facilitation and maintenance could potentially be built and well-regulated . It is of no wonder, therefore, that Indian Ocean has become of vital importance for Israel.
- Furthermore , having defined for themselves as being the three enemies that ought to be targeted in the wake on 9/11 attacks , it might be not wrong to say that the US, for its objective, might in the future , openly endorse support for Indian and Israel in whatever ways possible .
- Some weeks after 9/11 , this is how Washington post-columnist Jim Hoagland while writing in his article “a test of true allies” has described the future scenario:
  “India and Israel are the most vibrant democracies in a vast swath of countries from North Africa through the Himalayas that should now be seen as a single strategic region. Jerusalem and New Delhi are also end points of the US campaign [against terror]."
- India’s former Home Minister and deputy Prime Minister L. K. Advani echoed this view. On a FOX Television interview he stated: "terrorism insofar as we've seen it on 11th September ... has a common source, and that common source has described the United States, Israel and India as its three main enemies.” (Martain, 2003)

Implication's for the Security of Pakistan

- **Strengthening Defense Ties with India; Starting a proxy war in Pakistan – The Region of Balochistan and FATA**

  Terrorist are being provided training along the Pak-Afghan border controlled by the Taliban. In this area, 17 training camps, by Israel and India are in operation. Terrorists inside Pakistan, in particular in FATA, are now working together with Israel and India, carrying out their terrorist activities inside Pakistan. (Ishfaq, 2010) The Nation reports that to build up the propaganda warfare and misinformation among the security forces of Pakistan and NATO, India invited the Israeli activists to be involved in the anti-Pakistani activities. (The Nation, 18 June 2010).

- **Pakistan and the Muslim World in a Fix**

  It can be seen how after 9/11 India has toned down its condemnation for the crimes against humanity committed by Israel, and has avoided actively participating in UN resolutions against Israel.

  Israel statement stating "that an unwritten and abstract" axis with India and the United States has been created to combat international terrorism, has put the Muslim World and Pakistan in a colossal dilemma. On one hand, there are fears of US endorsement for Israel and Indian alliance, while on the other hand this alliance will allow India to take actions against radical extremism inside the territory of Pakistan. (Khan, A. p.145)

¤ **U.S.**

A unilateral declaration severely narrowed down the options of survival for Pakistan as a state.

Given to us were two options, from which the security forces of Pakistan choose to assist the army of the U.S. (Musharaf, 2004) Baluchistan, rich in oil and gas, an area of strategic value for the U.S. made us begin an internal war. It made us turn our soil into a battlefield between the people of Pakistan and the Army. (Raman, 2003) We even undertook a military action in 2003 against the "foreign elements" in tribal areas of South Waziristan, NWFP. The media of the West titled Pakistan as a "headquarter of global mail-order business in nuclear bomb technology with Libya, Iran and North

Korea as its known customers." We were rushed under the IAEA to give on our nuclear samples. (Dar, 2006) Sadly, today, after more than a decade has elapsed since the 9/11 attacks, all the propaganda, intervention and submission in this fight has engulfed almost every city and province of Pakistan killing thousands of soldiers and civilians.

U.S. – Pakistan Relations: More tactical than strategic

Farooq Hameed Khan writing in his article, "More Tactical than Strategic" writes about this relationship:

Our partnership with the U.S. and the drumbeat of strategic dialogue gets Pakistan nowhere near towards protecting its survival and integrity. Having known that about 150,000 officers of Pakistan had been in distress, there have been 8000 causalities in the 2007 FATA action, we still hear the tradition mantra of 'do more', take more actions and take further steps to battle militants .

Reserved for being called as 'the oldest and trusty ally,' U.S. diplomacy of pressure and coercion comes together with carrots and stick policy for Pakistan. More new sanctions, aid program perched and suspended and not to forget, violation of our territorial sovereignty by unilateral strikes in Waziristan are what one often hears about.

Recurring statements from the senior leadership in America, of threatening Pakistan with 'severe consequences,' only neutralizes the pro-aid program given by America and further rises the anti-American sentiments in Pakistan.

Ongoing dialogues, interactions between the US and Pakistani civilians and experts of security does not help much to build up for the trust deficit. When the top most leadership of Americans refuses to believe in Pakistan, regarding the issues of Bin Laden, Mullah Omar, or our actions against the Haqqani network in Pakistan, even after Pakistan is taking military action on a long standing demand by United States and Afghanistan on its own soil, with the Foreign Minister Spokesperson Tasnim Aslam saying that, "Pakistan has clearly stated that the operation in North Waziristan is against all terrorists." (News In-depth, 2014) what bigger level of trust do both states expect from one another?

U.S. - Double Standards

1. U.S. – India Nuclear Co-operation: For peaceful purposes?

On 1st October, 2008 an agreement called as the U.S.-Indian Civilian Nuclear Agreement or the Indo – US nuclear deal was signed. Esther Pan; Jayshree Bajoria in the article titled "The U.S.-India Nuclear Deal published in the Council of Foreign Relations mentions some discriminating features which include:

1. Non- prohibition of the U.S. nuclear trade with India. This agreement comes at a time while efforts, curbing nuclear proliferation and preventing states like North Korea and Iran from the acquisition of nuclear weapons, are being made. Also, nuclear states, including US, Britain, France and Russia are trying to limit their production. This deal is likely to lead to the escalation of an arms race in Asia.
2. In exchange against the deal , India can
   - Buy U.S. dual use nuclear technology
   - These would contain equipment that can enrich plutonium or reprocess uranium, which has the possibility to create materials that can be used in bomb making.
   - Imported fuel for nuclear weapons will also be given.
3. India for now has chosen to abide to the strict laws of the Nuclear Supplier Group (NSG). However, if it lifts these chosen restrictions, it can easily trade its weapons to other less trustworthy states.
4. The executive director of Nonproliferation Policy Education Center Henry Sokolski, states, "We are going to be sending, or allowing others to send, fresh fuel to India - including yellowcake and lightly enriched uranium -- that will free up Indian domestic sources of fuel to be solely dedicated to making many more bombs than they would otherwise have been able to make." Furthermore, according to a "Congressional Research Service report" There are no measures in this global partnership to restrain India's nuclear weapons program."
5. A CRS report details, "A significant question is how India, in the absence of full-scope safeguards, can provide adequate confidence that U.S. peaceful nuclear technology will not be diverted to nuclear weapons purposes."

6. There is no constraint onto the number of nuclear weapons that can be produced by India.
7. When there are more cost-effective ways (making the grid station more effective, making effective the coal industry and renewable resources) of upgrading India's energy and technology sector, why was there a need for this agreement, which was rushed and had no comprehensive overview. In the Non-Proliferation Review , William C. Potter , Director of the Center for Non-Proliferation Studies at the Monetary Institute of International Studies has written:
   "The agreement appears to have been formulated without a comprehensive high-level review of its potential impact on nonproliferation, the significant engagement of many of the government's most senior nonproliferation experts, or a clear plan for achieving its implementation .Indeed, it bears all the signs of a top-down administrative directive specifically designed to circumvent the interagency review process and to minimize input from any remnants of the traditional 'nonproliferation lobby." (The Washington Post, 2008).

2. <u>Preventing Pakistan from the Acquisition of a similar deal; US opposition to Pak-China Nuclear Cooperation</u>

On one hand, Hilary Clinton tells Pakistan to manage its water issues internally and solve the issue of Kashmir bilaterally with India. On issues of its own security and national interests U.S. interference is seen in Pakistan's internal and bilateral affairs. Just take the example regarding transfer of Chinese nuclear technology to Pakistan. Rewarding India with civilian nuclear technology, it opposes the bilateral Pakistan - China Nuclear Cooperation.

- As compared to the U.S. Strategic Partnership, Pakistan-China Strategic Partnership is one based on mutual trust and non-presence of pre-conditions. (Khan , 2010)

3. <u>On NPT; Pakistan and India are " very special friends"</u>

- The Under Secretary of States for Arms Control and International Security, Ellen Tauscher says, "We would make all countries to sign onto the NPT. We have a universality commitment, yes." The Obama administration, however, has refused to mark out countries, together with, India and Pakistan from signing the NPT, since they are "very special friends". Tauscher said,

"We have terrorist organizations and organized criminal groups and other bad actors who are looking to acquire nuclear technology, nuclear know-how and nuclear materials. Secondly, we have more states wanting to acquire nuclear weapons than we have in the last 15 years." (Hamid, 2010)

- In 2012, we heard US Defense Secretary, Leon Panetta, expressing fears about the nuclear weapons of Pakistan falling into the hands of the terrorists. Panetta said: "The great danger we have always feared is that if terrorism is not controlled in their country, than those nuclear weapons could fall into the wrong hands," (The Economic Times , 2012)
- In spite of the concerns, in 2014, US released its report on the nuclear material security, in which, Pakistan was recounted as the "most improved country" (The Express Tribune, 2014). In comparison to India and other nine nuclear-armed states, Pakistan was a state, which could protect well its nuclear materials.

4. Pakistan Nukes–Danger of falling into the wrong hands; India nukes–In hands of Hindu fundamentalist?

Every one blames the elements of Pakistan's intelligence ISI and their links with the Taliban, but no one has criticized or talked about the 'Indian Army-Hindu terrorist's nexus.'

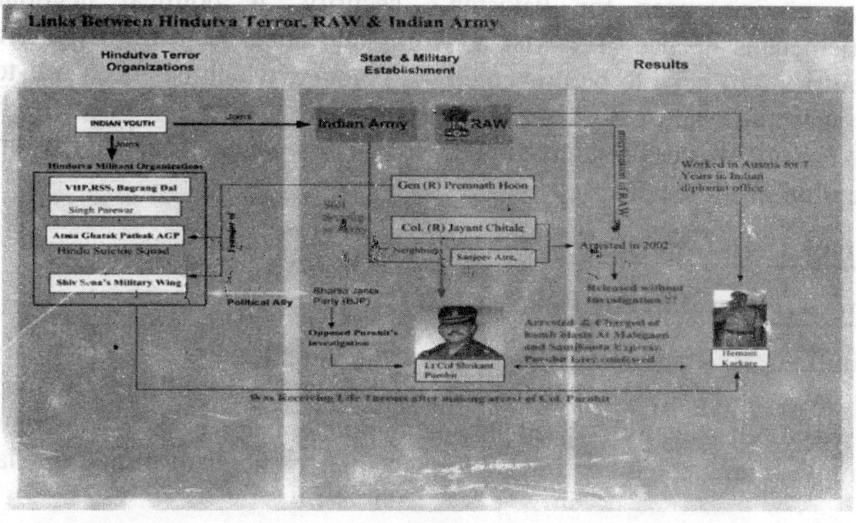

As written by Farzana Shah in her article, "Indian Army – Hindu terrorists' nexus: Sinister Fascist Plan", RAW is India's intelligence agency. In business of intelligence, details to the officials of the government are not essentially given, because the state is making headways on its objectives. The training programs and activities of terrorism inside and outside India cannot be done without an intelligence agency in India, which is RAW. It is strange to know that the 'world's biggest democracy is involved in destabilizing its neighbors.' Manipulated by Hindu fundamentalist organizations like RSS, Bajrang Dal and BJP, today there are fears that nuclear assets of India might fall into the hands of Hindu fundamentalist, who have made their place into the Indian army, where they have their agenda of 'Hindutva. ' (Asian Tribute, 2009)

¤ Afghanistan

Introduction

Afghanistan until today has not been able to construct itself as a stable and unified state. Our policy of "Afghanistan first" should be abandoned above the interest of Pakistan. (Sehgal, 2003a) What is left of it is a nation propped up by the West. There are, however, reasons for which Pakistan needs to remain concerned about Afghanistan, namely Indian presence, the buildup of peace across both sides of the Durand Line, and a proxy war against Pakistan in the Afghan territory. (Sehgal, 2003b)

According to Lodhi (2013), the post -2014 US withdrawal of troops will have consequences for Pakistan's security. A political negotiated settlement is needed. Terrorist networks and attacks on the neighboring countries will need to be prevented. A country, which has been war-torn for years, will add issues of the porous border security for Pakistan, thus making our western borders another frontier to which Pakistan's security attention would be diverted.

A Brief History

The Afghan Threat Perception

A buffer state contending against the external and internal authority, fight for power, Afghanistan has been a constant victim to all these practices. As has been said by Afghan President Rabbani, in the UN Millennium Summit after September 2001, "foreign interference had turned our land into a terrorist training camp, a center for drug smugglers and a base for spilling instability." (Singh, 2010)

Consequences - What followed the Cold War

- U.S. provided arms to Mujahedeen, when the Soviets had occupied Afghanistan.
- Consequently , in the vacuum created after the Soviet Withdrawal, Taliban regime was formed
- 9/11 incident
- U.S. Invasion - 2001 in Afghanistan
- Exposed to political, economic and religious manipulation by regional and extra regional powers (Singh, 2010)

After 9/11 - leaving with unfinished business

About the Afghan President, Hamid Karzai, in 2013, it was written:

> "That he was convinced that US President George W Bush's "war on terror" in the aftermath of the 9/11 attacks in 2001 should not have been fought in Afghan villages and homes. He believed that, "the war on terror was not fought with honesty and not fought genuinely; its consequences, thus, are being felt across the region." (BBC NEWS Asia , 2014)

An Unhealed Trust Deficit?–Factor's Straining Af-Pak Relations

A number of around ten policy research papers written by IR experts were compiled by CIDOB Barcelona Center for international affairs in 2012. Titled "Mapping the Sources of Tension and Interests of Regional Powers in Afghanistan and Pakistan," the papers present an in-depth study concerning the seeds of conflict between the global, regional and the neighboring state.

1. **Internal Stability in both the countries – Still in a fluctuation**

   There are issues of

   1. Governance
   2. Socioeconomic underdevelopment
   3. Ethnicity, sectarianism, radicalism, militancy
   4. Influx of migrant's across border (ethno-demographic changes)

5. Civil unrest in Afghanistan and Pakistan (fighting within its borders)
6. State institutions not being able to deal with these challenges

2. **The interests of the regional powers**

Regional and international actors identify Asia, being rich in energy resources and global trade, capable of meeting their energy, strategic and political needs. For international trade routes to be established, however, stability in both Pakistan and Afghanistan is important.

In addition to the interest of the regional powers of India, Iran, Russia, China and Saudi Arabia, U.S. post -2014 withdrawal from Afghanistan presents a situation in which either common interests can be successfully achieved or states could remain divided over divergent perceptions about them.

Interests of the Regional Actors

India	Iran	China	Russia	Saudi Arabia
• Eradication of the safe heavens of terrorists • Countering Pakistan's control in the region • Way in for access to raw materials for the needs of its population • Putting an end to Laskar-e -Tayyeba , which is used as a proxy by the ISI to act against India • Countering through Iran the Sunni militant's in Afghanistan, so that a save passage for commercial and energy routes, through Central Asia can be reachable	• Seeing, that a long-term military presence in Afghanistan by the US is not maintained • Enduring cooperation at the international level by the US • Making a contact with all the regional actor's working in Afghanistan • Avoiding, the rise of Taliban, which is inflectional with Saudi Arabia. • Counter-influencing the Shia Iran as opposed to Saudi-Wahhabi in Pakistan.	• Way in for raw material • China's mineral and infrastructural interest in Baluchistan • China-Pakistan relation's : Lever with US and Afghanistan • Gwadar Port: Countering Indian influence in the region and US influence in Central Asia • For internal security , Preventing the training of Uighur , inside areas of Pakistan	• Using the state as a buffer to prevent Islamic militancy • Preventing flow of narcotics into the Russia (Domestic concerns) • Presence in Central Asia to avoid US influence	• Ideological Interest's • Promoting itself as a Sunni Unifier • Preventing an Al-Qaeda Taliban Nexus • Using Afghanistan as a proxy battlefield between Tehran and Saudi Arabia • Promoting Saudi -Pak relation's (aid , investment and oil concessions) · • Pakistan-Keeping a

• In cooperation with the international community , India wants projection of its country as a regional power	• Apparent but ambiguous position with some regional partners (China and Pakistan) • Securing for routes of communication and energy. Spiraling economic ties with Afghanistan and India • Avoiding the Sunni terrorist from operating across border • Trying to maintain regime security in Afghanistan, so that its policies at home are not altered • Countering Pakistan's confidence with US and pro-Taliban (Within Pakistan) • Soft- power , regional influence , decreasing the Afghan refugees and taking measures to curb drug trafficking(domestic concerns)	like Fata • Propose an intend to establish military network in Pakistan , so that it's goods can be easily transported to Xinjiang province		status quo • Saudi-US privileged relations • Sphere of political influence -Influence with Mujahidin and Taliban . Role of a mediator in talks with Afghanistan and Pakistan • Investment and capitalization in both countries

Source : Compiled from the information presented in the CIDOB Policy Research Project, 2014.

Issues in Pakistan and Afghanistan – how it effects their relations?

Any act conducted from the territory of another state is solely the responsibility of the state and its institutions. In case of Pakistan and Afghanistan, one of the reasons for their mutual trust defiant and disturbed relations is that they have not been able to establish the writ of the state inside and that the situation of law and order in their own countries remains problematic.

Afghanistan – Issues of Governance	**Pakistan – State Indifferences**
International aid lost , due to corruption ; resulting in emergence of 'mafia' groups	Rise of the militant groups due to lack of basic services like justice and humanitarian assistance to the civil society
Weak Rule of Law • Weak system of justice • Lacking the power of enforcement = Weak rule of law	**Role of the Armed Forces** • Civil- Military gap • ISI is perceived by the neighboring countries, to be working upon an independent destabilizing agenda
Taliban and the Parallel Structures • Taliban's –Accepted among people as security guarantee's • West- Trying to empower local elites to marginalize the Taliban regime	**Non-State Actor's** • Criminal and sponsored Organizations • Insurgents and other institutions all take up the role of ' local government '
Centralized Political System-In Conflict • Constitution does not meet obligations • Non-Pashtuns against the political system	**Non –Extension of the Writ of Government** • Shelling across the border in areas like Kunar creates tensions with Afghanistan. • With Taliban insurgency active in the area , instability within the country , internal migration and displacement within the country adds to problems within the country
	Conflict and Ethnic Tensions in Balochistan Nationalist support from Afghanistan and India= Regional tensions Afghan Taliban , Shura of Quetta and networks of terrorist along Pakistan-Afghanistan Border =Tensions with Afghanistan ·

	Control for Ownership + Mineral Resource = Destabilizes interest of both state and regional actors like China
	Pakistan-Indian Tensions Pakistan's Accusation's- India uses the constable's in Afghanistan to train Bloch nationalist , creating instability in Balochistan Indian Accusation's- Pakistan to stop terrorist who plot against India With these tension's India therefore looks onto Afghanistan as an ' alternative route' for transport of trade and goods
	Pakistan-US Tensions US Accusation's – ISI uses the Afghan Taliban insurgents group Pakistan's Complain- Violations of its territorial sovereignty. Negative accusation of Haqqani network as a terrorist organization **US Uncertain Role in Afghanistan** Reluctance by the Taliban on negotiating with US US withdrawal from Afghanistan and the country , falling into the hands of the Taliban
	Fear's of the Regional Powers Return of Taliban Spread of Extremism Aversion towards the US and Pakistan's support for the Taliban
	Ethnic Conflict Pakhtuns would call for nationalist and separatist movements in Pakistan from Afghanistan Afghan National Army (ANA) being split upon ethnic - regional lines
Sectarian Conflict • Ethno-Sectarianism-Politically motivated violence in Pakistan • Sectarianism in Iranian Balochistan (Sunni Minority in Shia majority country), spills in Pakistan	**Sectarian Conflict** • Rare , but have happened (e.g. attack on Afghan Shia minority on Ashura Day) • Other regional actors or Al-Qaeda taking advantage of the circumstances can create tensions internally and

Balochistan. • Groups of Lashkar-e-Jhangvi and Sipah-e- Sahaba (Sunni Muslim militant organization) affiliation with Taliban and Al-Qaeda	between the relationship of Iran – Pakistan
• Drones attack in areas of Fata has given rise to radicalism Pew Research Poll – 74% =US is enemy **Growing Radicalism** • Private media routinely highlighting these extremist views • Counter- ideology to eradicate extremism is absent due to which people seek direction, security under the religious ideologies • Radicalism across Elites , lower and middle class	**Growing Radicalism** • Anti –Western sentiments have grown. Afghans see Nato/US as having failed to improve the lives of the people • Taliban established media message= Spreading hate messages
Militancy Afghan and Pakistan Taliban have different objectives, but call for international forces to leave and for Islam law to play some role in the state. Involved with them are terrorist groups from Central Asia exporting it to Pakistan , Afghanistan and within its own country	
The Pakistan Taliban Emerged , where the control of government was weak	**The Afghan Taliban** Cross border insurgency from Fata , KP and Balochistan

Source: Compiled from the information presented in the CIDOB Policy Research Project, 2014.

India:

Introduction

Both India and Pakistan have stated national security as a core objective for having declared themselves as states having nuclear weapons. This narrow concept of security has left both of the countries without having established for their states security policies, which could protect their respective nations' core values. These core values not only include the external threats, but also embrace anti-national and anti-social elements within their own countries. In democratic countries, India and Pakistan, due to lack of consensus between masses and leaders, the concept of core values has been abused to usage, for decision makers' personal gains. This has created discord between states, groups or individuals. In theory, India's external and defense policies seem to be holistic but in practice, it seems to be Pakistani-centric. However, its foreign policy takes into consideration other factors as well. (Budania, 2001) Pakistan's policy is developed on a similar pattern, presenting nothing more than a security dilemma, raising the possibility of an armed conflict between the two states.

Indo-Pak Animosity-Where does it begin from?

- **National Identity**

  'History is what makes nations what they are. It gives them their character, their institutions and their identity.' (The Economist , 1996). In her paper, "Cultural exclusivism to cooperation: Prospects for Indo-Pak harmony", Meneesha Tikekkar reflects upon the educational construct of cultural identity. Briefing, she mentions that

  Hindutva and Islamisation have used this history based on information and evidences as a tool of cultural socialization. Blurring the lines between reinterpretations and rewriting , history, mythology and memory, the textbooks read in schools are given out according to historian 'nationalists' and 'official'. After independence, India played the game of manipulation, projecting cleverly and later, deliberately selective history. (p, 239)

- **Monoliths Culture**

  Both Hindutvavadis and Islamist are threatened by the complexities of culture and plurality. Cultural Monoliths have tried to entail single and absolute version of the culture. It invites violent reaction resulting in an inward looking mindset, protecting the narrowed traditions. As Krishna Kumar says, "We remain unprepared to recognize that Pakistan and India are each other's domestic problem, psychologically serving as highly destructive material for nation-building. We have used each other to justify and embrace the suicidal ideology of militarist nationalism." (Kumar, op.cit. p. 240.)

- **<u>India and Pakistan; Opposite Doctrinal Variables</u>**

  In the paper titled "Indo-Pak peace process: military & political dimension", Dr. Hussian (2004) says that ideology or values and belief system of the India and Pakistan are different from one another. This explains their threat perceptions towards each another. Quoted by S.M. Burke, "Centuries of dedication to such diametrically opposed systems as Islam and Hinduism could not but nurture an utterly different outlook on the outside world among their respective followers." (p. 44)

- **<u>Irredentism: The Kashmir Dispute</u>**

  In his book, "India's National Security Dilemma: The Pakistan Factor and India's Policy Response", Rajpal Budani highlights the case of Kashmir issues as following:

  "In the post WWII period, one of the longest unresolved disputes between India and Pakistan has been the issue of Kashmir. The existence of religious commitment and separatism across the borders of both the countries are a threat for the integrity of the states. Kashmir, by Pakistan has been claimed as its part of the territory, due to it being a state, where majority of the Muslims have lived. India on the other hand , negating anti-irredentism (ethnic or religious statehood; non-military approach), puts forward its argument that if religion was a bidding factor, there would be no Bangladesh, no identity crisis and no ethno- nationalism in Pakistan."

<u>Kashmir Issue: becoming more complex</u>

1. Opposing ideologies in the sub-continent
2. Strategic location of Kashmir
3. Internationalization of the issue
4. Participation of external powers
5. According to Gupta (1996) Ideological diversions (Congress – A secular , strong , stable India ; Muslim League –Areas in the northwest and east of India = Separate , strong , a vital strategic state) (p. 440)
6. A part of Kashmir , negotiated with China by Pakistan in 1963 has put off any leeway for holding a plebe site
7. India's constitutional adjustments of making Kashmir an integral part of its territory. (Lok Sabha Debates , 1963)
8. Wars over this issue

9. The psycho-strategic dimension
10. Insurgency and psychological warfare in Kashmir
11. India's status quoits policy (national-security specific)
12. Core values of survival for both India and Pakistan
13. India's no acceptance to UN resolution or third party resolution
14. Building upon public opinion for highlighting Pakistan - sponsored terrorism (Budani, 2001)

- **The Siachen Issue**

The largest, only source of fresh water for the Indian sub-continent, and a source of water for irrigating the lands of Punjab, Siachen is another ground of battlefield between India and Pakistan, since 1984, Siachen glacier has exemplified mountain warfare. Non-demarcation of the cease-fire line is the root cause of the conflict.

When it comes to the resolution of the conflict, both the states put have divergent points of view over the question of how effectively can troops from both sides can be withdrawn.

India's Position on the Issue	Pakistan's Position on the issue
" Demarcation " or "authentication" regarding the troops position , for if after the region has been demilitarized any military action taken by Pakistan can be responded to	Mutual withdrawal of troops to the pre-1984 position , for any demarcation can in the future become an advantage for India, to hold talks for resolving the issue

(Kaswan, 2013)

With issues of climate change and rise and fall of water level around the world, India and Pakistan are among those countries that are moving towards becoming water-stressed. With the armies of the two countries still occupying and battling in the area, in the years ahead, as the glaciers melts, the nature of the conflict is going to see a massive transformation. Rather than being entrapped mentally and physically while keeping their positions on this front, both the countries should seek to resolve their socio-economic problems. Since life in South Asia, sustains on water, there can be problems of migration, poverty, and malnutrition. (Chaudhry, 2012)

Water Dispute – India's unilateral construction of the dam

India is making the most out of its geostrategic position. By construction of the barrage and additional water capacity India will be providing water flow for its domestic and economic purposes. The unilateral construction of the Wullar Barrage, build in the disputed region of Jammu and Kashmir not only violates the Indus Water Treaty, but also affect Pakistan's hydropower generation capacity. (Mohan, R.C, 2004)

The Paradoxes

Retd Brig. Feroz Hassan Khan presenting his paper on the "Evolving stability in South Asia," explains the existence of paradoxes between India and Pakistan as following:

- The Stability- Instability Paradox

  Glenn Snyder "Stability-Instability Paradox" states that two adversaries, having nuclear weapons may make both states become involved in "limited" military adventures against each other. This may be done so, as long as the vital interests of the states are not hurt . When this paradox is applied to states like India and Pakistan, Pakistan's endeavor of moving for a High Intensity Conflict (HIC) in Kargil, 1999 and Indian concept of "limited warfare" observed during the 2001-2002 compound crisis can be seen as examples in which nuclear weapons can become the cause of instability in the region of South Asia. With deterrence optimist assuring a "restrained crisis", pessimist are of the view that inherent doctrines, policies and practices of both countries give birth to new unresolved issues, with the previous problems still unresolved. (p. 175)

- The Dependence-Independence Paradox

  On having exercised their independent nuclear capabilities, India and Pakistan have faced challenges against regimes like non-proliferation and maintenance of nuclear weapons, but in practice, both India and Pakistan are not deterred by each other. Rather, this effective deterrence is brought in by a third factor- intervention by the US. Since 1990's what has hindered the crisis from turning into a war, is our dependence on the US intervention and diplomacy. For every crisis if US diplomacy is to be used to stabilize the region, then it reflects the dismal of deterrence stability. Being depended on Western powers, both the states, while pursuing their version of national security objectives, are in fact making themselves more dependent on the West. (p. 175)

- The Unintentional Predicament

For an "inadvertent nuclear escalation" between India and Pakistan, Professor Barry Posen identifying three scenarios states that if in an asymmetric environment, with Pakistan lacking a geophysical depth and conventional military power over India, any interplay of conventional and strategic forces between the two states begins, then it might force a nuclear strike. (Posen, 1991)

- Nuclear Stability in South Asia-evolving threats and strategies

In an international three day conference, titled 'Nuclear Deterrence and Emerging Dynamics in South Asia' organized by the South Asian Strategic Stability Institute (SASSI) ,developing threats and Pakistan's dependence on nuclear deterrence was discussed . The theme of the conference has been elaborated below.

The evolution of new concepts of limited warfare, the Indian Military Cold Start Doctrine, proactive acquisition of Ballistic Missile Defense System (BMS) along with increase in the conventional defense system by India i.e.$ 43 billion against Pakistan's little over than $6 billion, the arms race is going to increase the problems of stability, security and prospects for the region of South Asia.

The director general of SASSI, Dr. Maria Sultan said:

"Concepts such as the Cold Start Doctrine and Proactive Operations are based upon the notion of pre-emptive strike and calls for rapid deployment of Integrated Battle Groups. Cold Start Doctrine joint with massive militarisation force has the capability to increase the level of an arms race, hence raising the level of minimum deterrence stability in the region. Developments in the neighborhood compelled Pakistan to increase its dependence on nuclear deterrence and by drafting strategies in view of evolving threats. NASR, Short Range Ballistic Missile system is one of the key elements of a 'Full Spectrum Deterrence (FSD)' strategy. The Full Spectrum Deterrence strategy is a response measure against the evolving threats, mainly emanating from neighboring nuclear-armed India."

Also according to General Sikandar Afzal, "The discriminatory policies relating to nuclear cooperation pursued by some major powers was creating insecurity and imbalance and Pakistan has been compelled to take a stand against nuclear exceptionalism, selectivity and discrimination."

While for others including Masood Khan Khattak, Indian modernization of its military was to operationalize the cold start doctrine and achieve its regional hegemony.

Relating to US-Indian Nuclear Deal, Hou Hongyue remarked that China did not consider India as a nuclear threat, but US –India nuclear deal and development of ballistic missile systems were a issues of grave importance for international stability. (The News, 2014)

- **India's Maritime Security Policy**

India seeks to control the crossways in the Indian Ocean, which connect the busiest strategic routes of maritime communication. Vice Admiral Sangram Singh Byce has described its emphasis for them in the following words:

"The Indian Ocean encompasses about one-fifth of the world's sea area and the Indian Peninsula just two thousand kilometers into the sea, bringing about 50 percent of the Indian Ocean within a 1000-mile arc ascribed from the Indian Territory. India's dominant position in the middle of the Indian Ocean and astride some of the busiest strategic Sea Lanes of Communication imposes on us huge operational responsibility to ensure security of both choke points and SLOCs so that there is free and unhindered flow of international trade."

Objectives of India in the Indian Ocean

- Spreading influence – This would be done through trade , investment , strategic partnership along with diplomacy
- Securing for its country, energy assets for economic development through its relations with Africa, the Middle East and Southeast Asia
- For the years to come, it would like to see its country positioned as a dominant power
- Preventing Chinese foothold in the region

- Being in command of choke points in the Indian Ocean. To secure this objective, building security relations with littoral states like Singapore, Mauritius and Oman would be maintained. (Ranasingh, 2011)

Implication's for Pakistan

With China and India, growing toward the next economic giants in Asia, securing energy needs for their increasing populations is a vital concern. Accordingly, the US naval presence in the region has become a concern for China. According to Mathur, it is observed that a system of alliance between India-US has been strengthened to counter China-Pakistan alliance. India fears "China's string of pearls" strategy, through its port construction in its neighbors. According to Muhammad Ali, India also plans to inflict a naval blockade on Karachi. (Eurasia Review, 2014)

Under these circumstances and South Asia's nuclear dynamics, Pakistan would, in the future, have to make hard choices for lying down strategies to secure its own energy routes, land and national interests.

Pakistan is becoming a victim to external aggression and internal challenges. If internal stability is not achieved, our borders and environment will become exposed to any threat that works against our ideological values, territorial integrity and survival.

In contrast to the developed state, in developing states like Pakistan, internal vulnerabilities are present in a greater degree. Achieving the socio-economic and political goals in such an environment is therefore, challengeable. (Budania, p. 25)

Thus, the next chapter discusses the internal security challenges facing Pakistan.

Suggestions and Recommendations

As goes the Chinese proverb, "When you are in a common boat, cross the river peacefully together." (Wright, 2009). Thus, it is imperative to fight the enemy from within first, and then align with those across the borders, to eradicate the enemy outside.

Pakistan's Mainstream Nationalist Narrative

In the book "Muslim Zion: Pakistan as a Political Idea" Devji (2014) narrates and clarifies the contradictions viewed for the creation of Pakistan. Establishing a framework, he has reviewed points of thought, which can help Pakistan protect and establish and progress upon its core ideological values without debating further more on the states purpose and its creation.

- Idea played an important role in shaping the history of Pakistan.
- "Muslim', became a meaning, through which to define a community that needed representation."
- Islam principally defines Pakistan's identity.
- The means to guarantee the status and positions of the Muslim within India, in the post-colonial period was resolved by the definition of Muslims as a 'nation'.
- Creation of a Muslim state, which was in huge numbers to be called a minority in India, resolved the communal discord. India was left at liberty, to build its own national identity and democratic process
- Iqbal and many others had argued on the defining Muslim nationalism laying its foundation not only on communal observance but also on beliefs reaching subcontinent.
- When one talks of Muslim solidarity that needs to be established within and beyond its borders, it means 'negation of alternative forms of identity' and 'of adherence to a specific vision of what is meant to be a Muslim.'
- However, 'Pakistan is a nation without history, rejecting its past through the pursuit of a universalistic, de-territorialized Muslim identity as a means through which to construct its future.' (with reference to provincial harmon)
- As General Kayani has been counseling, "If you want to fight with history … it's your choice to do so. No individual or institution has the monopoly to decide what is right or wrong in defining the ultimate national interest," the general had warned that, "As a nation, we are passing through a defining phase … Weakening of the institutions and trying to assume more than one's due role will set us back."(Qaiser , 2014)
- Pakistan needs to take the initiative. As has been stated in the book, "The Prince."

  "It must be remembered that there is nothing more difficult to plan, more doubtful of success, nor more dangerous to manage than the creation of a new system. For the initiator has the enmity of all who would profit by the preservation of the old institutions and merely lukewarm defenders in those who would gain by the new ones." In this regard, Turkey for us sets an example. A Pildat Discussion paper published in 2012 on "National Security Council: A debate on institutions and processes for decision making on security issues" mentions how through good governance, support of the people and its duty in the international community, NSC balanced the civilian-military relations.

- The 21st century may be dominated by headlines of killings of the innocent and devastating nations, but the future is in the hands of those nations, who "decide to work with one another, not against each other." (Islam, 2014)

❖ Af-Pak Relations – Re-energizing the on-again , off-again relations: Policy of Pursued Engagement

Journalist Graeme Smith in his book, "Lessons from the Afghan war" states that the rise of Talibans in the Afghanistan-Pakistan border has been a factor of two failed miscalculated strategies (Sethna, 2014)

1. American Miscalculations - Air strike escalation
2. Pakistan's home grown extremism

With Pakistan, having rejected the 'terrorist ideology,' today it has shown the outside word its resolve to make its contribution to fight the global war on terrorism.

One of the major problems that Afghanistan and Pakistan have is that of terrorism and cross border terrorist activity. Only joint ventures in terms of military and economic progress can help in lessening the intensity of these conflicts.

US General , Joseph Dunford states that , if Pakistan and Afghanistan are non-cooperative, there is 'significant risk' of return of Al-Qeada to the region, (Iqbal , 2014) meaning more use of drone strikes and revival of terrorist safe heavens in both countries.

Strategy	**Grand Strategy**
• **Border Security Forces (BSF)** A border paramilitary force to guard the lands of both countries during peacetime and prevent transnational crimes and better	' Pursued Engagement' • Looking into prospects of Joint Social and Economic Cooperation, Isolation for both countries is not an option.

monitoring and registration of the inflow of migrants from across the borders • **<u>State Migration Laws and Migrant Centers</u>** o Freedom of Movement o Family Unity o Procedural guarantees in areas of detention and expulsion o Consular access o Rules for irregular and labor migrants	• **<u>Joint Military Co-operation</u>** o Bilateral military exercises
• **<u>Security Pact</u>** <u>Zero Tolerance Strategy for Terrorism</u> Taking genuine steps to control extremism in our lands by granting authorities the right to take firm actions against terrorist • <u>Non-Violation of Territorial Integrity and Sovereignty</u> Not to allow their respective territories to be used by 'foreign forces' to attack each other's territory.	
• **<u>Contact Lines –Strengthening Communication Networks</u>** In times of crisis, to exchange views from across the borders for verification and authentication of allegations. Both of the countries need to understand that many of the regional and global actors have their objectives which they want to pursue via the lands of Afghanistan and any attack on its territory can be falsely alleged on Pakistan, which in reality can be an act of state-sponsored terrorism or non-state actors.	• **<u>Social Co-operation</u>** <u>Health;</u> Joint campaigns for eradication of polio and other health related problems. <u>Agriculture;</u> Agriculture for both of the countries is their livelihood and to sustain this livelihood both the countries need water. Policy of <u>non-confrontation</u> should be adopted by both the countries and problems of boundary water should be organized and monitored.

• **Re-emergence of Taliban and Reconciliation?** As Clinton (2014) said, "A surge of Military forces without any effort on the political side is not likely to succeed. So infact it is a combined strategy that makes a great deal of sense. " One of the obstacles for direct talks of Karzai with Taliban's, have been elements from Pakistan's ISI. They did not want Karzai to reach upon a separate deal with Taliban and want to prevent potential Indian influence in the country. So, if Pakistan takes the initiative , and regional actors agree , here is what can every regional actor benefit from **Regional Security Pact – Pakistan , China , Iran , Afghanistan and Russia : Reconciliation with Taliban** **"Preventing Afghanistan from internal destabilization "** Common Interests of Regional Actors • Minimized role of US in Afghanistan • Preventing militancy infiltration and activities across borders • Preventing internal destabilization in Afghanistan Major Interests of Afghanistan • Stabilized Afghanistan • Preventing re-emergence of extremist elements • Guaranteeing protection to the majority Pashtuns and minority ethnic groups If reconciliation is to be	• **Economic Co-operation** Pakistan – The key that unlocks the gateways for trade corridors o The Waziristan Corridor Connecting trade hubs of Afghanistan-Kabul and Kandhar , through the Central Trade Corridors (Central Trade Corridors) o The RSRBCG Strategy (Replacing, Shrinking, Reducing , Building , Creating and Generating) ▪ Replacing- From Roads to Highways ▪ Shrinking- Reducing distances from Northern-Western Pakistan and South-Eastern Afghanistan ▪ Reducing- Building integral network of roads from into Waziristan ▪ Building- Communication gateway between Afghanistan and Pakistan ▪ Creating- Health and education facilities ▪ Generating- Economic Opportunities (Khan)

considered , then o Taliban 's (those who want peace and not imposition of any ideology) and the locals are to be given " internal protection " against Al-Qeada and extremist elements , by deployment of peace mission forces formed by those who sign the security pact . o In return, the Taliban's will only fight against the terrorist and will not support terrorist campaigns inside Afghanistan. o In case of re-emergence of terrorist groups , for protection of their respective borders and preservation of stability in Afghanistan , these states peace mission forces should be comprised of regional forces.	
	• <u>**Cultural and Educational Exchange**</u> Programs, for eradicating elements of extremism and separatism. • <u>**Development of Education Facilities**</u> With knowledge will come understanding, awareness and eradication of extremism.

Source: Author's Own

❖ Pakistan and India - Survivors in South Asia

Either Pakistan and India can continue to live in status quo, or by accepting each other's existence as a reality, we can make South Asia, a continent of peace for our people.

As Krishna Kumar says, "we remain unprepared to recognize that Pakistan and India are each other's domestic problem, psychologically serving as highly destructive material for nation-building. We have used each other to justify and embrace the suicidal ideology of militarist nationalism." (Tikekar, 2004)

The 21st Century; A New Century; New Security Challenges

The Non-Security Challenges

Where to begin?

Non-state actors making use of non-traditional weapons
HIV/Aids which has no borders and is a threat for social order
Raising population lacking food security
Melting glaciers and risks of floods , drought , famine , disease and migration
Poverty and illiteracy

Setting pre-conditions for talks- Not an option	Setting before ourselves ' pre-conditions' for talks will not help both the countries move forward, for whenever both the countries continue to pursue 'peaceful resolution of issues,' external designs and internal interventions prevent that to happen. As said by Tasnim Aslam (2014) "You cannot have dialogue and at the same time impose pre-condition" (Dawn , 2014)
Ceasefire Violations – A tit for tat will take both the relations of the countries to the levels	If both states keep engagement in an "action-reaction syndrome," over violations like that of ceasefire, how much level of trust are they trying to develop from such actions?
Deterrence Stability – A Reality	Between nuclear-armed states, like India and Pakistan, deterrence = stability, as said by Ali (2014), "Islamabad has developed an indigenous, affordable, survival and credible nuclear deterrent to survive in an insecure, changing lonely environment."

	Where there is division in interests and both of the states are adversaries, concerns of security on both sides will continue to increase. In comparison to U.S.-Soviet Cold war deterrence stability strategy , what is of concern for both countries is the fact that they have geographical proximity and are involved in countermoves for more nuclear weapons , stockpiling of materials e.tc. Among the five ways presented by Michael Krepon to avoid the pitfalls of nuclear-armed competition, the best option that can be applied to lower tensions between India and Pakistan is
Propaganda Warfare-A Reality	• As said by ex investigator India officer Satish (2014), "India itself was behind Parliament in Delhi and Mumbai terror attacks. Both terror attacks were set up with the objective of strengthening the counter-terror legislation." After the attacks, India started allegations against intelligence and agencies of Pakistan. • Another reality as expressed by Chuck Hagel is, "India for some time has always used Afghanistan as a second front, and India has over the years financed problems for Pakistan on that side of the border. (The Times of India , 2013) In an interview with stating Indian involvement in sustaining the insurgency in Balochistan , Musharaff on an India TV channel it was said , " the ISI did not train Kashmiri militants, but admits they got shelter in Pakistan. There is no question of the army, intelligence or the ISI creating Mujahideen groups, training them and sending them in... Hizbul Mujahideen group was created in your (Indian) part of Kashmir and then they ran away from there after military atrocities. (The Express Tribute , 2010)
Kashmir Issue- Conflict Avoidance is not Conflict Resolution	The more avoidance of Kashmir issue, the more problematic and complex will it be to resolve. Irrespective of whatever the

	claims are by India and Pakistan, both have to learn from the history of WWII, where Germany and France fought over the territory of Alsace/ Lorraine. The problem has to be resolved through dialogue. Schofield (2014) in her paper , " Kashmir : The way forward " has presented the following options for resolution: • Independence for the whole of Kashmir • LoC , being turned into international border • Bifurcation of the state • Maintenance of the status quo • Regional referendum with collective will backed by third party facilitation , for guidance In the interest of both the states, she presents the fifth option as the most suitable option. Let the conflict not be drawn to a point, where Kashmiri's fight for justice be seen as crimes of terrorism or nationalism.
If Not Peacefully – Both countries would not hesitate to start a war; The hostile attitudes continue to persist	Just before the meetings that were going to be held in August 25th, India's new army chief Gen Dalbir Singh Suhag took charge and on an ending note, this is how he created an atmosphere for an upcoming meeting "We dare you start a war, and you will be delivered the same fate again," To this Pakistan's naval officer bleated, "We are fighting in Waziristan to protect you, but if you want to fight us you are welcome to try crossing the border." (Dawn , 2014)

If we do not stop this tit for-tat game, it will not be long before we see our future, much like present North-South Korean rivalry.

Source: Author's Own

Confidence-building and nuclear risk-reduction agreements, implicit or explicit so that the effect of destabilization and nuclear competition is reduced (Dawn , 2014)	→	Both India and Pakistan , not only can avoid the dangers of nuclear war , but can also prevent wars similar to those of the past , which in turn will prevent third party intervention and imposition of international sanctions

❖ Pakistan and US – "Working on Complained Relation"

In politics, trust is the only factor, which can strengthen relations between states. As for US and Pakistan, they would have to go a long way to mend their relations.

Violation of Territorial Integrity and Sovereignty	The Use of Drones	Air Defense Identification Zone
	CIA operatives: Not being part of the U.S. armed forces, these individuals under international law are participants in hostilities, just like those whom they target. Under International Humanitarian Law(IHL) , therefore: 1. If these combats are targeted it could be viewed as a legitimate attack on an unlawful combats. 2. Under domestic law, the combat loosing immunity could be prosecuted for murder, by a state. Henceforth , not only are drones being used by a covert entity , not trained in the laws of war , but also, these strikes	For lessening of anti-American sentiments, protection of territorial sovereignty, integrity, protection of national honor , Pakistan should identify an 'air defense zone' under which 'permission is granted in authorization, by the government of Pakistan to undertake strikes.' Since Pakistan , has now committed itself to fight 'All terrorist,' it can continue to fight its war of survival more effectively and precautiously.

	impede proper mechanism for transparency and accountability. In case of U.S. drone strikes inside Pakistan, due to the mixing between the civilian and combat population, weapons proliferation in the tribal areas, it is not easy to tell apart when does an action of a non-combatant becoming a participant of an armed group. U.S. needs to clarify its bases for attack or provide for what standard of proof is applied. (Center for Research and Security Studies)	
Still worried about Pakistan being an ally?	According to Congressman Dana Rohrabacher, a campaigner of an independent Balochistan, "US funds were being used to persecute religious and ethnic minorities in Pakistan... Pakistan is not an ally...they are allied with terrorist elements and even more dangerous adversary, Communist China." (Dawn , 2014)	Policy of Harmonization Pakistan should form procedures under which it is to exchange with the government of the US a committee report that verifies the ' development outcomes ' which have been achieved by US funds and to address donor–government partnership and problems of governance.
Spying by NASA-Activity contrary to the friendly relations	According to Khan, as part of the massive sying to surveillance program National Security Agency (NSA) has been given permission by the US to 'spy' on Pakistan, along with 193 other countries. (Dawn , 2014)	Any activity of this nature, will strain the U.S.-Pakistan trust deficit, for spying on Pakistani government departments, organizations and individuals is "violation of international law". Rules to protect the digital infrastructure and information should be pursued.

Source: Author's Own

❖ Iran and Saudi Arabia Proxy War in Pakistan – "Protecting Our Core Values"

Resolving the ideological divide	Iran-Pakistan and Saudi Arabia all are representatives of the same religion, Islam. All the three countries need to maintain internal stability and prevent regional instability. Funding institutions inside another country to spread ideological difference is not a wise option. Instead, these funds should be used for constructive purpose i.e. to say to build and uphold seminars and activities (e.g. calligraphy) of the two communities under 'one umbrella.' Use the platform to help the ordinary people understand the true meaning of religion and the values associated with them. This will eventually decrease the level of usage of hate speeches and killings, which spreads negative messages, across the community and world. Exploiting the masses, for political gains will only invite external actors to gain advantage by strengthening more divisions.

Source: Author's Own

References

1. Ahmad, M. (2003) Shi'i Political Activism in Pakistan. *Studies in Contemporary Islam, 5*(1-2), p. 64.
2. Anderson, P. (2012). *Why Partition?* London Review of Books, 34(14), 11-19.
3. Abass, H. (2010, September 22). *Shiism and Sectarian Conflict in Pakistan Identity Politics, Iranian Influence, and Tit-for-Tat Violence.* From https://www.ctc.usma.edu/wp-content/uploads/2011/05/CTC-OP-Abbas-21-September.pdf Retrieved June 14, 2014.
4. Ali, S. M. (2014). Why Small nuclear weapons will not disarm? OPED *Eursia Review, news & analysis.*
5. Ali , M.S. (2014 , August 23)*Why small nuclear powers will not disarm ? - Oped* From http://www.eurasiareview.com/23082014-small-nuclear-powers-will-disarm-oped/
6. Retrieved August 1, 2014.
7. Ahmed , T. (2014 , January 11) 2014 report : Pakistan ' most improved ' in nuclear security , India not so .
8. Budania , R. (2001) . *Indian National Security Dilemma: The Pakistan Factor and India's Policy Response*, M.L. Gidwani, Indus Publishing Company.
9. By The Newspaper's Correspondent (2014; June 23) Congress rejects move to stop US funding for Pakistan. *Dawn*.
10. B. Raman, ‚Al-Qaed a's New Trojan Horse,' Outlook India, 10 November 2007.
11. For details, see Hassan Abbas, ‚Defining the Punjabi Taliban Network.,' CTC Sentinel 2, no. 4
12. (April 2009), 1-4.
13. By the Correspondent (2014, August 2) India media upstages army in hostility towards Pakistan. *Dawn*.
14. Cohen, S. P. (2004) .*The idea of Pakistan*. Brookings Inst Press.
15. Correspondent, O. (2014, May 15). Developments in India forced Pakistan to increase dependence on nuclear deterrence. *The News*
16. Chaudhry, S. (2012, March 12) The Siachen issue. *The Express Tribune*.
17. DeSilva-Ranasinghe, S. (2011). India's Strategic Objective in the Indian Ocean Region Future Direction International.
18. Dudkevitch, M. (2003, December 8).US force in Iraq adopt IDF Strategy. *Jerusalem*
19. *Post.*

20. Darby, B.M. (2013, August 9) *Different Political Trajectories: India and Pakistan*From http://www.e-ir.info/2013/08/09/different-political-trajectories-india-and-pakistan/
21. Retrieved 4 September, 2014
22. Dar, H. (2006). The US Invasion of Iraq (2003) and Diplomacy of Pakistan. *Pakitan Vision, 7*(2).
23. Esther P, Jayshree B, Staff W (September 4, 2008) The U.S. India Nuclear Deal. *The Washington Post.*
24. For details on the Zia era, see Hassan Abbas, Pakistan's Drift into Extremism: Allah, the Army and America's War on Terror (New York: M. E. Sharpe, 2005), 89-132.
25. Gen Beg, A.M. (2014, July 21) Proxy war and politics in Pakistan. *The Nation.*
26. Gupta, S. (1996). *Kashmir: A Study in India Pakistan Relation* New Delhi: Asia Publishing House.
27. Johnsons, L. J., Kartchner, M. K., &Larson , A.J. (Eds.) ( ). *Strategic Culture and Weapons of Mass Destruction : Culturally Based Insights into Comparative National Security Policymaking* .
28. A national interest is a public good of concern to all or most Americans; vital national interest is one, which they are wining to expend blood and treasure to defend. National interests usually combine security and material concerns, on the one hand, and moral and ethical concerns, on the other. See, Samuel Huntington, "The Erosion of American National Interests," Foreign Affairs, Vol. 76, no. 5 (September-October 1997)
29. Halim, A.A. (2003, November 23). The Indo-Israel military relationship and its
30. Impact on the Arab world. *TheMonitors.*
31. Hafeez, N. (2012) *Evolution of national security structure in Pakistan.* From http://www.issi.org.pk/region-publications.php?region=South%20Asia&start=31
32. Retrieved 2September, 2014.
33. Hussain, D. R. (2004). *India - Pakistan peace process: military & political dimensions.* Paper presented at the Prospects of Peace, Stability and Prosperity in South Asia Islamabad
34. Ishfaq, S. (2010, August 29).Indo-Israel nexus is a threat to regional peace. Pakistan.
35. *Daily News.* ·
36. Indian Army-Hindu terrorists' nexus: Sinister Fascist plans. (2009, February 19). *Asian Tribute.*
37. Islam, S. (2014, July 19). Connectivity is the new buzzword. *The Dawn*, p.13.
38. Iqbal, A. (2014, July 19) Al Qaeda may return, warns US general. *Dawn.*
39. Javid, H. (2014, August 3) Cover Story :Muslim Zion: Pakistan as a Political Idea. *Dawn.*
40. Khan, A.Z. (2011). Development in Indo-Israel Defence Relations Since 9/11
41. Khan, H.F. More tactical than strategic! (2010, July 27). *The News.*
42. Kumar, K. op.cit

43. Khan, B. C. E. M. Conflict Transformation by Pakistan Army: The Central Trade Corridor (CTC). From https://www.ispr.gov.pk/front/main.asp?o=t-article&id=95. Retrieved July 1 , 2014.

44. Kaswan, P. (2013, 4 February) *Siachin Conflict: Background and Current Situation.* From

45. http://www.simplydecoded.com/2013/02/04/siachin-conflict-background-and-current-situation/

46. Retrieved July 8 , 2014.

47. Khan, A. I. (2014, July 18) Pakistan asks India not to set conditions for talks. *Dawn.*

48. Krepon, M. (2014, August 3) A nuclear myth. *Dawn.*

49. Khan, A. I. (2014, July 4) Foreign Office Protests to US over spying by NSA. *Dawn.*

50. Katja, R (2007). Sectarianism in Pakistan: A Destructive Way of Dealing with Difference, Brief No. 2, Pakistan Security Research Unit, University of Bradford.

51. Lodhi, D. M. (2012, December 18). Our security calculus, *The News.*

52. Lokh Sabha Debates (November 1963). *22-27*

53. Legality of U.S. Drone Strikes in Pakistan. Retrieved July, 4, 2014, from http://crss.pk/downloads/Reports/Special-Posts/Legality-of-US-Drone-Strikes-in-Pakistan.pdf

54. Mapping the Sources of Tension and the Interests of Regional Powers in Afghanistan and Pakistan. Retrieved 4 September , 2014, fromwww.cidob.org/.../INFORME+...

55. Martin, S. India and Israel Forge a Solid Strategic Alliance. (2003, February 28). *The*

56. *Jerusalem Post.*

57. Mohan , R.C , 2004 *Cultural Exclusivism to cooperation: Prospects for Indo-Park Harmony*. Paper presentenced at an international seminar organized by the Institute of Regional Studies, Islamabad.

58. Mathur, S. China, Indian Ocean & the Nuclear Deterrence: The Destabilizing Effect: Institute of Peace and Conflict Studies.

59. Ms. Elizabeth L. Stone, D. C. P. T., and Dr. Peter R. Lavoy. (2005). *Comparative Strategic Culture.* Paper presented at the Conference organized by the Center for Contemporary Conflict, U.S. Naval Postgraduate School for the Advanced Systems and Concepts Office of theU.S. Defense Threat Reduction Agency., Monterey, California,

60. Mumtaz, A. (2003). Shi'i Political Activism in Pakistan,.*Studies in Contemporary Islam,* 5(1-2), 64.

61. Nasser, N. (2007). Indian-Israeli Ties Could Neutralize Delhi's Palestinian Policy.

62. Nuclear Treaties. (23, April 2010) *Dawn.*

63. Niccolo Machiavelli Quotes. Retrieved 4 September, 2014, from http://quotemeanings.net/niccolo-machiavelli-author-5254.html

64. Oldenburg, P. (2010). *India, Pakistan, and Democracy: Solving the puzzle of divergent paths*. Routledge.Milton Park, Abingdon, Oxon: Routledge.

65. Of late there is growing consciousness among the scholars about the damage done to history by zealous historians. For critique of history writing in Pakistan, see the writings of K.K.Aziz, and Mubarak Ali and for criticism of the school textbook, Pervez Hoodboy, A.H.Nayyar, and Rubina Saigol in particular. For history writing in India the most recent work are by Krishna Kumar and Vinay Lal.

66. P. R. Kumaraswamy. (2005, March 11) A post-Arafat shift. *Online Asian Times.*

67. Petras, P. J. (2011). Puppets in Revolt: Afghanistan, Iraq, Pakistan and the United States.

68. Pinault, D. (1993). *The Shiites: Ritual and Popular Piety in a Muslim Community* (p.59) New York: Palgrave Macmillan.

69. Pant, V.P. (2004) India, Israel Partnership: Convergence and Constraints. *The Middle East Review of International Affairs, (8)* 4.

70. Posen, R.B. (1991) *Inadvertent Escalation: Conventional War and Nuclear Escalation*. Ithaca: Cornell University Press, 1991)

71. Qaiser, A. (2014, January 19) *The Garrison State's Civil-Military Divide, Part 3: Pakistan's National Security and Intelligence Apparatus* From https://www.cdainstitute.ca/fr/blog/entry/the-garrison-state-s-civil-military-divide-part-3-pakistan-s-national-security-and-intelligence-apparatus

72. Retrieved 4 September 2014.

73. Raman, B. (2007, November 10). Al Qaeda's New Trojan Horse. *Outlook India.*

74. Riikonen, K. (2007, March 1) *Sectarianism in Pakistan: A Destructive Way of Dealing with Difference,* Fromhttps://www.ctc.usma.edu/wp-content/uploads/2011/05/CTC-OP-Abbas-21-September.pdfRetrieved June 14, 2014.

75. (Retd) Brig. Khan , H.F. (2004) Cultural Exclusivism to cooperation: Prospects for Indo-Park Harmony. Collection of paper presentenced at an international seminar organized by the Institute of Regional Studies, Islamabad.

76. Sahgal, A. (2003, September 19). Israel and India join forces. *Asia Times.*

77. Sehgal, I. (2003a). Afghanistan First" Policy". *Defence Journal, Vol .7* (1).

78. Sehgal, I. (2003b). The New Great Game *Defence Journal, Vol.7* (1).

79. Singh, RSN (2010) iDR INDIAN DEFENCE REVIEW. Retrieved from http://www.indiandefencereview.com/spotlights/afghanistans-strategic-culture-and-threat-perception/#

80. Sethna, R. (2014, July 6) Cover Story: Lessons from the Afghan war. *Dawn*

81. Schofield, V. (2004, June 7-9) Cultural Exclusivism to cooperation: Prospects for Indo-Park Harmony. Collection of paper presentenced at an international seminar organized by the Institute of Regional Studies, Islamabad.

82. Singh, R. (2010). Afghanistan's strategic culture and threat perception.

83. Sectarian War: Pakistan's Sunni-Shia violence and Its Links to the Middle East,' Event Summary, Woodrow Wilson International Center for Scholars, 2 May 2007. See also Khalid Hasan

84. Tahir, M. (2014, July 27) News Analysis: Pakistan's major anti-Taliban Waziristan offensive revives tension with Afghanistan. *News. In - depth.*

85. Tikekar, M (2004, June 7-9) Cultural Exclusivism to cooperation: Prospects for Indo-Park Harmony. Collection of paper presentenced at an international seminar organized by the Institute of Regional Studies, Islamabad.

86. Unwritten, Abstract US-India-Israel Axis to Fight Terror. (2003, September 11). *Indian Express.*

87. Wright, T. (2009, December 18) *Clinton at Copenhagen: 'We are in a common boat.'* From http://www.3news.co.nz/world/clinton-at-copenhagen-we-are-in-a-common-boat-2009121807 Retrieved 4 September 2014.

88. Afghan President Karzai dismisses ' al-Qaeda comeback. ' (2014, June 18) *.BBC NEWS ASIA.*

89. India to spend $3.5bn on military modernization. (2014, July 20) *Dawn.*

90. Pakistan's nuclear weapons ay fall into the hands of the terrorists: Leon Panetta. (2012 , August 12) *The Economic Time* .

91. (1996, 21 December) p. 74. *The Economist.*

92. India supporting militancy in Balochistan: Musharraf. (August 20, 2014). *The Express Tribute.*

93. India financed problems for Pak in Afghanistan, says US defence secretary nominee Chuck Hagel. (August 25, 2014) *.The Times of India.*

94. 9/11: Pakistan's Security Concern and Policy Options. *South Asian Studies 26*(1), 131-151.

95. 9/ 11 commission report, final report of the national commission on terrorist attack upon the US. (2004).

CHAPTER NO. 4

PAKISTAN FOREIGN POLICY CONSTRAINTS

Introduction

Foreign Policy is face of a state in international community. Not only National interest is achieved through it but also decides the status of a state. States with clear and long term foreign policy is developed or on the way to become one. Well established foreign policy makes a state a super power while foreign policy designed for self-interests makes the state a weight in scales of international politics. Every state wanted to be dominant in global arena, to preserve its national interest but without string foreign policy it remained a futile struggle. The words written in documents are needed to be implemented according to their worth and a state with powerful foreign policy tools can convert those theoretical ideas into reality.

Pakistan has been in the dilemma of changing the words into reality. Security dilemma has been the core of its foreign policy since inception. Foreign policy makers of Pakistan have to focus on security issues for survival of state. Since then the foreign policy has been revolving around security issues more or less. Along with security constraint Pakistan has been facing numerous hurdles in foreign policy making process. Preference to Self-interest over national interest, external pressures, and in-efficient diplomats and recently the domestic circumstances are few of them.

Inefficiency in building a strong foreign policy has caused Pakistan a great damage. The state image at international level is at decay, trust deficiency with other states is a serious issue for Pakistan. Internal disharmony among national institutes has made the policy making more complicated and much weaker. Weak and self-absorbed leadership makes the foreign policy a document for personal use. Options opted by our leadership in post 9/11 world has become the reason of rise of internal threats along with already existing external issues making national security more complex.

Foreign policy which supposed to be a supporting element for national security has converted into a constraint in case of Pakistan. The policy makers have remained unable to develop a clear national security policy because of the reasons mentioned above. To develop a strong foreign policy to support national security policy, Pakistan government has to overcome these issues. These constraints are mostly self-created and Pakistan has the ability to resolve them by learning the lesson from history and mistakes.

Foreign Policy Constraints

"Our foreign policy is one of friendliness and goodwill towards the nations of the world. We do not cherish aggressive designs against any country or nation. We believe in the principle of honesty and fair-play in national and international dealings and are prepared to make our utmost contribution to the promotion of peace and prosperity among the nations of the world. Pakistan will never be found lacking in extending its material and moral support to the oppressed and suppressed peoples of the world, and in upholding the principles of the United Nations Charter." (Quid-e-Azam, 1948)

- **Trust Deficit**

Deficiency in trust among global community is the major concern for Pakistan. USA and Pakistan has enjoyed close relations in different phases and both are front line allies in war against terrorism. Despite this an American writer Bruce Riedel wrote about Pakistan "Pakistan is the most dangerous state" (Riedel, 2012, p.1). This statement by a writer of Pakistan's hypothetical ally depicts the status of Pakistan in international community. The image of Pakistan has been on the descending slope especially in post 9/11 era. Despite the suffering due to war on terror which proved to be "War for Terror" for Pakistan, the state holds the image of terrorist state in global community (Hilali, 2013).

Major reason of this trust deficit is internal security situation of Pakistan. The non-state actors brought up by Pakistan have turned rogue to their creators. These elements has started to target with in borders of Pakistan and making a negative impact on state image. Now Pakistan is being taken as somewhat an anarchical entity which is threat to regional and global peace. Although the socio-economic structure is at its worse and the state has the burden of 40,000 causalities still international community is not ready to trust Pakistan.

Foreign policy and weak leadership of Pakistan are the significant reasons behind this trust deficit. Wrong choices were being opted at wrong time in history. Decisions being took were short sighted and not in national interest. The aftershocks of those decisions are long lasting and destructive. It will take Pakistan decades to come out of these faulty policies. Pakistan government and establishment have to work around the clock to bring country on right track. Strong and clear foreign policy based on National interests and preserving national security is the critical tool for future benefit and prosperity of Pakistan (Askari, 2004)

Trust Deficit with India

Both Immediate Neighbors are at Logger's Head since their inception. The territorial, economic and cultural differences have increases the relational cleavage between them. Three wars, numerous border clashes and Kashmir issue have made the situation worse. Both states are not ready to compromise on their stances (stance on what). Geo-strategically and economically both state are vital for region as well as global politics. The rivalry between these neighboring states left a deep impact on regional and international security scenario. The major reason of enmity is the trust deficit between both neighboring states.

Kashmir issue is the main matters of conflict between Pakistan and India. They fought first was over the issue in 1948 just after a year of their independence. Both states claimed the Kashmir as their integral part and both are not ready to step down from the Tangible issue. Pakistan supports the ongoing Freedom Movement in Kashmir and India has been blaming Pakistan regarding the cross border movement into Kashmir. Both have their own side of picture hence claiming it the right side. In this whole scenario the existence of trust is quite of question especially when Indian government accused Pakistan that the state is using its land for harnessing terrorism against India. Likewise Pakistan is of the view that India is sponsoring and nurturing the terrorism in Pakistan especially in Balochistan and FATA region by funding Baloch Insurgent Groups in Balochistan and TTP in FATA (PILDAT, 2010)

Pakistan and India are located in Critical Economic zones and region. According to reports both states contain the 92 % of South Asian GDP, 85% of South Asian Population and 80% of the South Asian Surface area but the percentage of the trade between them is merely 20%. (Raihan, De, 2013). Lack of trust which leads to increased gap in communication has brought the situation to this stage. Both states need to understand their strategic and economic significance for the stable South Asian region (Husain, 2012).

Rocky Relations with Afghanistan

Both states have rocky and historically stuck relations since 1947. The border "Durand Line" is the core of conflict. Afghanistan never recognized the 1500 Km Long line which sowed the seed of rivalry between both neighboring states. The Afghan establishment has always claimed that the major part of Pakistan tribal areas and Balochistan is belonged to Afghanistan. Both states are strategically interlinked and without a stable Pakistan, Afghanistan cannot be peaceful and vice versa. The historical blame game and absence of trust worthy policies has increased the relational gap between both entities (Rehmani, 2006)

Pakistan's direct inclement in Afghanistan started during American "Afghan Jihad" in which Pakistan decided to be a front ally. The Taliban phenomena took birth under the maternal watchful eye of Pakistani agencies and US government. After the USSR disintegration, the Taliban Mujahedeen became the useless for US so Pakistan was left to deal with them solely. In 1990s Pakistan was one of very few states who had diplomatic relations with Taliban regime. Remaining international community started to show concerns on Taliban regime and their methods of rule. The start of 21st century and 9/11 incident changed the global and regional scenario for Pakistan as well as for Afghanistan. (Bajoria, 2009)

The "Freedom Fighters" became "Terrorists" overnight. Pakistan again chose to be a front ally of US in its new war in Afghanistan by taking a U-turn in its foreign policy regarding Taliban. After Taliban regime was toppled, a new democratic government was established in Afghanistan by US. Despite Pakistan's role in War against terrorism, the friendly and trust worthy relations were not established between both states. Karzai government started to raise the issue of Durand Line and blaming Pakistan for cross border terrorism (Husain, 2011) In Tribal areas FATA along with the Afghan borders, some factions of Taliban established new wing under the name of TTP. Afghanistan claim that TTP conduct terrorist activities in Afghanistan and hide in the shelters in Pakistan. Meanwhile Pakistan has the view that Afghanistan is funding these factions with help of India and the Afghan authorities provide the shelters to TTP terrorists who are involved in terrorist activities in Pakistan (AIAS, 2007)

The loose and long border is hub of the problems as per official statuses thousands of people daily cross the Pakistan-Afghanistan border. Only at Torkham point twenty to twenty five thousand individuals cross the border on daily bases with only few hundred of them having complete travel documents. Thousands use the crossing points at Chaman, Ghulam Khan and Angoor Ada. Along with that there are number of unchecked points in

Tira, Chitral, Kurram, Dir, Mohmand and Bajor. A strong and secured border and close diplomatic ties can be the vital tools for bridging up the gap between these two neighboring states (Khan, 2012)

Roller Coaster ties with USA

After its creation Pakistan chose US to be its first Power full ally. Relations between them have passed through many phases and ups and downs. USA is being blamed for using Pakistan in dire need and abandoning it after wards. The incidents of arms embargo in 1965, ban on nuclear activities of Pakistan, US tilt to India, leaving Taliban in hands of Pakistan in post-Cold War era and despite its sacrifices in war against terrorism, operations against Osama Bin laden, Pakistan is still standing on the grounds of suspicion and non-trust for international community including USA (Hasnat, 2011)

USA got itself engaged in direct and indirect wars in South Asian Region and both times Pakistan chose to be part of its clash strategy. Firstly the Taliban were created by US, for US and then were abandoned. At second time the war was launched to eradicate the same Taliban again by US, for US. Pakistan was a front line actor on both occasions. The aftermaths of these wars directly hurt Pakistan and its society specifically and majority of Pakistani people blame US for this whole havoc. According to a recent survey 80 percent of Pakistani citizens think that Al-Qaeda is standing against US and 57 percent supports it and 52 percent put the blame of US for their societal, economical destruction in post 9-11 era (Fair, 2009)

The mistrust and cleavage in relations between has increased at large scale after 9/11, Although US has spent $11.2 Billion between 2002-2008 Fiscal years and another $1.2 Billion in 2009 on Pakistan. Along with this US has allocated "Kerry-Lugar Bill" of $ 7.5 under the "Enhanced Partnership with Pakistan Act 2009". This bill was severely criticized by the Pakistani masses as they took as an American step to control their economy even at micro level (Hasnat, 2011)

Strong diplomatic relations and the impact of relation at masses level will fill this gap between both states. US has to approach to a common Pakistani for trust gain in place of its political elite and Pakistan has to use its diplomats, embassies and Consulates in US to communicate with US establishment and citizens to enhance its image and trust gaining (Cohen, 2011)

Tension among Institutes

Internal disharmony instates of Pakistan has bring the state to this position. From very start the harmony lacked between them. The state faced dearth of leadership after Quid-e-Azam which made the situation critical. The vacuum was created and to fill that gap, institutes started to take over each other and it is happening till now. Two main institutes Establishment and political government are at logger's head almost in every phase of history of Pakistan. Both institutes blamed each other for current situation of the state. Political elite is of the view that Establishment especially Military has not allowed them to run country according to their policies while Establishment tool political leaders as in-competent (Rizvi, 2010)

This disharmony has resulted in short term policies adoption, incomplete projects and chaotic institutional performances. Society and political leadership both are clueless and confused (CRSS, 2011). This internal clash has resulted in disoriented foreign policy and major constraint in Making of National security policy. A well described foreign policy will paved the way for establishment of policy for national security. This institutional disharmony has brought Pakistan to its current place and taking it further in decay. All institutes needs to work in their legal boundaries to take Pakistan towards a better position in future. Internal strength will make Pakistan externally strong.

2. Trouble among Institutions

Pakistan politics and institutional structure has remained in turmoil more or less throughout its short history. Military coups, Establishment created politicians, PCO Judges, incompetency of political figures and established bureaucracy and confused democracy are some of major factors of institutional instability in Pakistan. The weak institutions and disharmony among them leads to weak and disillusioned foreign policy and no national security policy. The institutions' circle of influence is unclear hence they are overlapping one another making internal system of Pakistan more fragile (Hussain, 1999)

According to a survey conducted in 1997, there are four variables in political-institutes of a state which are strongly co-related. These variables are:

- Variables covering the government quality and performance of the institute i.e. rate of corruption

- Protection of basic and property rights and the imposition of its rules
- Type of government i.e. state is under democracy, dictatorship or monarchy etc.
- The societal structure of the state i.e. the level of income, gap in socio-economic graph and ethno-religious composition of the society

If a study is conducted under these variables or the aspects of institutions in Pakistan, a clear picture will be come up regarding the current situation of institutions in the state. Every passed government in Pakistan has experimented and utilized these institutions according to their self-interest. The National interest is being ignored by all political and military rulers (Alesina, 1997)

Pakistan has tastes almost every type of government in very short span of time. The incompetent political leadership has provided the chance to military generals to throw them over and ruled most of the time in Pakistan. From very start military was the only stable and functional institute of Pakistan hence forth it not only controlled the borders but also the foreign, economic and poetical fronts of the state. The left over years for ruling is being taken by Political parties and in place of making democracy thrive they have made it more confusing, unclear and self-centered.

Type of Democracy and Institutes in Pakistan

Pakistan like many third world states has mixed democracy where both parliament and Constitution has the supreme power in system. All the institutions get their power of work from written constitution. According to the Constitution Superior judicial body (Supreme Court) has the power to nullify any law passed by the government as Supreme Court of Pakistan has declared laws passed by parliament null and void but this power of Supreme Court does not make constitution supreme as parliament has the power to amend the constitution (Khan 2012)

Pakistan has been under military regimes most of its part since 1947. The military dictators have established the parliaments and National, provincial assemblies with their own puppets. The elected civil governments have never had a "Legislative Exit" in past until the last government from the 2008-2013. Own Mistakes and incompetency of politicians have led the circumstances to military coups repeatedly over the years. The politicians created by Military rulers and institutions handled by them have little chance to work efficiently. The rapid and continued changing in the regimes and governments in Pakistan are most negatively affected for the state and political institutions. (Malik, 2003)

Judiciary and Military Regimes

Along with the Parliament the second most affected institution affected be changing regimes is judiciary. The Judges and judiciary were used as personal tool in number of cases on number of occasions. The PCO judges issued the orders and decisions that had changed the course of Pakistan's History. Judiciary had remained a playground of rulers especially of military rulers. A Scholar of Legal Studies has stated that Pakistani courts have established for the protectors of constitution but over the years same courts have somehow justify and accommodate the constitutional amendments by the rulers. They did that because it was necessary for their own survival.

According to an International Crisis Group (ICG) Report of 2004 "The judiciary not only failed to check extra-constitutional regime change, but also endorsed and abetted the consolidation of illegally gained power". This has not only violated the constitution but also disturbed and destroyed the law making process and performance of the institutions to a large extent.

In 2007, Chief Justice of Pakistan Iftikhar Choudhry and other Supreme Court Justices refused to resign on the orders of Musharraf. This step by them brought the judiciary to a new turn where it was no more under the pressure of any ruler. Lawyer movement followed by that refusal resulted in the re-allocation of Supreme Court Judges. Chief Justice of Supreme Court started to use "Sou-Moto" action and wanted more investigation and accountability of numerous cases, bureaucrats, police and politicians. This was an "unprecedented in Pakistan Law and Political History" stated by Hassan Abbas a Fellow Researcher at Harvard Kennedy School of State (Bojaria, 2008)

The last democratic government completed its five legitimate years for the very first time. The Chief Justice's Sou-Moto initiative have not been taken so positively by democratic government even and it was called "Crossing the institutional Limit" and Parliament has stated that Supreme

Court is overlapping the judicial boundaries and creating hindrances in Parliament working process. Nonetheless the system is on the move and institutions are learning to work in their limits and constituency. The coming years will make the system more strong and better working institutions will emerge.

3. Territorial Issues

Geographical location of Pakistan is very critical. Along with that the state is located in one of most turmoil targeted and vital economically important region. The regional is under consistent instability for last numerous decades. Pakistan has the troubled neighbors like Afghanistan and also has its own borders issues since 1947. At its independence the foremost threat Pakistan faced was the territorial threat from its immediate neighbors. Eastern border was in-secured by India while Afghanistan was threatening at Western borders. To tackle this security threat Pakistan has to engage itself in security allies and pacts (Burke, Ziring, 1990)

The Eastern border of Pakistan with India has remained a permanent security threat. The Indian hegemonic designs make the border of both states unstable. Three full fledge wars and many borders clashes are the proofs of territorial issues. Kashmir unresolved issuc has make the border clashed much common between these neighboring states (Asia Report, 2012) This territorial conflict has not only affected the security aspect bit also has took the lion's share of Pakistan economy. Border with Afghanistan also called Durand Line is also an unstable threat to Pakistan. At time of Pakistan independence the neighboring Islamic state started to demand for piece of Land along with borders. Afghanistan has the point of view that this part of land belonged to them which was taken by British Empire and since 9/11 the Western border has become more matter of concern of Pakistan (Husain, Latif, 2012)

Both Eastern and Western borders conflicts has clutched Pakistan in an "Iron Fist" from both sides. The state has to fight on two fronts at a time which taking huge amount of the economy and man power. The deficiency of trust between these regional powers is a major reason of these clashes. Pakistan needs to develop its internal strength and relations on equal footings with its neighbors to tackle this issue.

Territorial Conflicts with India

Bone of Contention: Kashmir Issue

Kashmir is the bone of contention between both states since their independence from British Rule. None of them is ready to move from their stance or to change their point of view regarding the issue. Pakistan called Kashmir "Aorta of Pakistan" while India declares it as "Integral part of India". These statements show the grave importance of this issue for both states. This issue has caused wars, border disputes and hostility not only between them but also in South Asia region. Both parties have signed numerous agreements and numbers of UN Resolutions are being passed for the resolution but still the issue is unresolved (Rahman, 1996)

In 1947, Kashmir was one of 564 Princely states which were given right by British Kingdom to decide their fate and annexation either with Pakistan or India. Kashmir being a Muslim Majority (77 percent according to Census of 1941) and geographically attached to Pakistan was likely to become the part of Pakistan. The issue was raised by its Hindu ruler who wanted to annex Kashmir with India. The Muslim Majority in the region revolted against it and along with them their fellow tribesmen in Pakistan also revolted and they started to capture the towns. Maharaja requested India to interfere on the condition that he will hand over Kashmir valley to India, India signed a treaty with him and sent its troops to valley, Pakistan was not ready to accept this scenario and henceforth both newly free states fought their war over this issue in 1948 (Hussain, 2009)

India presented the case in UN for international intervention. On January 1949, Pakistan and India signed an agreement on Kashmir issue. In 1965 both states again went to war over Kashmir issue. In 1999 another was about to broke out between them again in the context of Kashmir. In 1998 the Kargil conflict was again for gaining Kashmir control. Because of this unresolved issue the political, social and economic atmosphere remains hostile almost for whole time. Kashmir issue is the core of conflicts and without resolving this conflict both states can never move forward in their bilateral relations.

Siachen Conflict

The Times (July 1989) declared it "War on top of the World". The Siachen Conflict is a by-product of Kashmir issue and has the ability to lead to a major war. The Kargil war was interlinked with the Siachen issue which was a limited war. Armies of both states are fighting in the most Cruel weather conditions in the world as Siachen is also known as "Third Pole". The Armies are not only fighting with each other but also with harsh weather conditions. These weather conditions are more dangerous to both armies but India has faced more causalities in this conflict.

Siachen Conflict was surprisingly started by India in a sudden Military Operation n 1984 completely violating the Karachi Agreement 1949 and Shimla Summit 1972. Since then Both Armies are engaged in cross border firing and fighting with the ruthless weather. The Issue of Siachen started with the ambiguous border agreements between Pakistan and India in which the both entities did not gave the clear boundary lines for this area. Now they both present their own side of picture and as this is also a tangible territorial issue, no one is ready to show flexibility on it. (Ahmad, 2006)

Sir Creek

The issue of Sir Creek is one of the eight major unresolved issues between Pakistan and India. The issue is based in the region of Sindh which is Southern Part of Pakistan and Indian region of Rajhistan. The disputed territory is almost 96 Km long water strip in the marshes of Runn of Koch of Indus River. Although like all other issues between Pakistan and India, this issue has not resolved but this both states have move forward to resolution in this conflict. At 14th SAARC meeting, Foreign Ministers of both states met and a press release was issue that both states are close to resolve this issue and they will issue a common map of the area (Khan, 2007)

Durand Line Conflict with Afghanistan

The 1500 KM long border line between Pakistan and Afghanistan known as Durand Line is the region of hostility. Kabul never accepted this border line and claimed the some areas of FATA along with the border as their own. Afghani 'Amir' in 1893 Abdur Rehman Khan agreed to this border line with British rule but since 1947 no political or monarchical regime of Afghanistan recognized it. Afghanistan wanted a free region for 'Pashtuns' of Pakistan that region where as Pakistan had the view that Afghanistan also had to free its Border 'Pashtun' region which Afghan authorities refused to accept. The issue is in deadlock and both sides are firm on their stance (Meyerle, 2013)

This Border area became more critical during "Afghan Jihad" in Zia's era when almost 3 million refugees moved into Pakistan from Afghanistan. Those refuges settled in those border areas and have the cross border connections. In 2011 when Pakistan decided to go against Taliban Regime in War of Terrorism, huge number of Taliban militants established their shelters in Border regions which lead to new factions of Taliban in Pakistan. In War for Terrorism the region became more unstable and dangerous. The newly American-led democratic government in Afghanistan started to blame Pakistan for Cross border terrorism in its territory while Pakistan has same accusations for Afghanistan (Baig, 2010)

The unstable border has further destabilized the Pakistan-Afghanistan relations. The area has not only becomes the hiding heaven for Militants and Taliban but also a Money Bridge for smugglers. According to the Afghanistan Opium Survey Report by UN Drug and Crime Cell, 98 percent of global opium is being harvested in Afghanistan. This opium is flooded in Pakistan through the loose border and smuggled to the world. This whole scenario is not only affecting Pakistan politically but also socially because of the flood of these drugs into society (Bajoria, 2009)

Despite the lack of trust and hostility between Islamabad and Kabul, both sides have conducted many rounds of talks to resolve this issue. The agreement and trust building is necessary for both sides to establish a stable border and close ties. In 2013 Pakistan and Afghanistan signed a strategic agreement which was first time that both parties were starting strategic dialogues for long lasting partnership in the region. This agreement will be a positive impact on relations and regional security atmosphere of South Asia (Ramin, Siddique, 2012)

4. Trade Security

Trade and economy are the national priorities in this globalized world. The policies especially foreign policies are designed for gaining the economic benefits. Even the recent wars are being fought for economic benefits. Geographical and Strategic location of Pakistan makes it a significant trade choke point. Important trade beneficial routes of Pakistan are Gwadar port, Trade for Afghanistan to Arabian Sea and Trade route to Central Asia for India. These all aspects make Pakistan a vital state in economic and trade market but the state has not took required benefits from its capabilities.

Weak Foreign policy is again the reason of this deficiency. The Past governments remained unable to develop policies to secure these trade benefits and to use them from economic prosperity. Priority to Self Interest over National interest is the main reason of this economical decay. The current government needs to improve this trade security and not only improve its economy but also earn from providing routes to neighboring states.

Pakistan's Regional Trade Opportunities

Pakistan is Located in one of the most critical region of the globe both economically and strategically. The Indian Ocean, Arabian Sea and to be operational Gwadar port are few examples. Pakistan shares the borders with four states, out of these four, three states e.g. China, India and Iran are significant actors of international economic market. Pakistan is the most suitable sea trade route for land locked states like Afghanistan and Central Asian States (CARs). Being an agricultural state, Pakistan is among the major exporters of agricultural goods, cotton yarn and leather goods. Gwadar port can be a golden economical gate for Pakistan once it's operational as its only deep sea port in the region. Through this port, Pakistan can become a major partner in trade of Indian Ocean. Despite these all positive opportunities, Economy and trade of Pakistan is in negative way.

The Political turmoil, weak economic structure and weak institutions from the very inception never let Pakistan to become a strong economy. The hostile India in East and not so friendly Afghanistan in West took the policy focus of Pakistan towards Security. Relations with Iran also remained more of hostile for last three decades. The security issues took the lion s share of Pakistani Budget. The adventurism by rulers of Pakistan also makes the state as a security focused state. Its regional neighbors like China and India has prioritized economy over security while Pakistan is still stuck in the security dilemma. These circumstances has made the trade balance of Pakistan in Negative as according to the data of Federal bureau of Statistics, the trade balance of Pakistan is as follows:

Value of Foreign Trade

(Million Rupees)

Year	Month	Exports	Re-Exports	Imports	Re-Imports	Balance of Trade
2008-09		1,383,717.5	21,927.3	2,723,569.9	1,642.9	-1,319,568.0
2009-10		1,617,457.6	23,470.1	2,910,975.3	14.3	-1,270,061.9
2010-11		2,120,846.7	30,576.3	3,455,285.6	2,072.6	-1,305,935.2
2011-12		2,110,605.5	5 18,570.2	4,009,093.0	3,131.7	-1,883,049.0
2012-13		2,366,477.8	9,946.3	4,349,879.5	-	-1,973,455.4

Source: Federal Bureau of Statistics of Pakistan. Retrieved from: http://www.pbs.gov.pk/sites/default/files/tables/14.1.pdf

China is the major and prominent trade partner of Pakistan in the region. To keep a balance in the region, China has opted to stand at side of Pakistan against the US-India Nexus. The trade between Pakistan and China took a critical turn in 1979, after the construction of all-weather Karakorum High Way. Although the rate of trade through that route is very small but still it got vital place in trade of both states. The trade between both states has increased in last decade and the reason is Free Trade agreement (FTA) between both states. In 2010 China was the 2nd largest importing state for Pakistan and fourth largest market for its exports. The alarming thing in this trade relationship is that Pakistan is exporting its raw materials, agricultural goods which not only create the dearth of these products in local market but also the imports from China are more cheaper which are directly damaging the local small industry (Hamid, Hayat, 2012)

Iran and Pakistan have faced numerous ups and downs in their relations. The economic ties are at the low level significantly because of the religious and ideological differences. Up till 2008 Pakistan's share in Iranian exports is merely 1.13% of its total exports and the share in imports are just 1.38% which shows the economic gap between both neighboring states (FPCCI, 2009) The gas pipeline started in 1990s for meeting the increasing energy needs of Pakistan, the gas pipeline through Afghanistan is also in doubtful conditions as Pakistan has not completed its side of pipeline and the involvement of US in this project has also made this project uncertain. Although the former government of Pakistan has ratified the project in 2013 but current government has not yet ready took the steps for its completion. In his recent visit to Iran, Pakistani Prime Minister Nawaz Sharif has showed the consent to increase the trade value up to $5Billion. The close economic ties between both states will not only result in the economic beneficial but also will help in regional stability (Tribune, 2014, May 11)

The neighbor with enmity India is emerging as a fast growing economy. Both Pakistan and India have never become economically close despite the trade opportunities. Little Diplomacy and communication gap has widened the gulf. India although has declared Pakistan as Most Favorite Nation (MFN) since 1996 but practically very few steps were taken. Pakistan recent past governments also has taken positive steps for trade with India like giving India the status of MFN in 2011. Current Government in Pakistan is also in favor of economic trades with India and showed its interest in projects like Pak-India-Afghanistan Trade route. Good trade relations between Pakistan and India interestingly highly depend on the security issues. A small unrest at borders or incident like Mumbai attacks will totally de-rail the trade partnership between them (Rizvi, 2012)

Economic gate to Central Asian States

Central Asian States are the hub of new economic great game of the globe. The region is full of natural and energy resources and world want to get them. Russia as the regional super power wants it share in it while USA as a global master wants to capture these resources. Whereas rapidly growing economies China and India are also trying to get energy-economic benefits. Pakistan's critical economic and energy sector situation direly demands the new gateways of energy and markets for economy. The Central Asian States can be that resource full place. In Post-Soviet Central Asian region, Pakistan can use the religious and cultural card tactfully

Former president of Pakistan General Musharraf during his visit in 2006 to the Central Asia stated: "Pakistan provides the natural link between the SCO states to connect the Eurasian heartland with the Arabian Sea and South Asia. We offer the critical overland routes and connectivity for mutually beneficial trade and energy transactions intra-regionally and inter-regionally". Despite the little success Pakistan is trying to increase the trade relations with CARs. Three options are critical to enhance the close trade relations between Pakistan and Central Asian States:

- Extension of Karakorum Highway (KKH) to Tajikistan and Kyrgyzstan

(b) The reconstruction of road linking Peshawar with Termez in Uzbekistan involving a distance of 880 km and that provides easy access to the remaining four Central Asian Republics.

(c) Land route linking Chaman with Kushka in Turkmenistan involving a distance of 1170km.

Gwadar Port is another significant opportunity for bridging up the trade gap between two entities. Gwadar is located at cross roads of Middle East, Central Asia and South Western Province of China. The land locked location of Central Asian States and Pakistan's strategically vital Gwadar port can be the corridor for export of Central Asian resources to the other globe whereas it also can increase the regional trade relations (Khan, 2007)

Iran's Energy Resources for Pakistan

Iran and Pakistan have faced numerous ups and downs in their relations. The economic ties are at the low level significantly because of the religious and ideological differences. Up till 2008 Pakistan's share in Iranian exports is merely 1.13% of its total exports and the share in imports are just 1.38% which shows the economic gap between both neighboring states. The gas pipeline started in 1990s for meeting the increasing energy needs of Pakistan, the gas pipeline through Afghanistan is also in doubtful conditions as Pakistan has not completed its side of pipeline and the involvement of US in this project has also made this project uncertain. Although the former government of Pakistan has ratified the project in 2013 but current government has not yet ready took the steps for its completion. In his recent visit to Iran, Pakistani Prime Minister Nawaz Sharif has showed the consent to increase the trade value up to $5Billion. The close economic ties between both states will not only result in the economic beneficial but also will help in regional stability.

Conclusion

Foreign Policy of Pakistan is one of the weakest trades of its states craft. The security dilemma, insincere leadership and nonexistent future vision of that leader ship have brought Pakistan to current situation. A clear and National interest based foreign policy can be success key of a state in international community. In current circumstances of turmoil, Pakistan needs a foreign policy which not only tackles the external threats but also meet the internal challenges. The state is on the double edged sword of threats. Incompetent politicians and diplomats have made the foreign policy more confused and short termed. The constraints like trust deficit has harmed thee state both at regional and global level. Despite its frontier role in international affairs Pakistan is one of the states on which not many global actors are ready to trust. This makes it a state with almost no strong ally and trust worthy partner. The territorial issues made the situation worse by creating the hostility with its neighbors and these conflicts alive and growing since 1947. These conflicts took Pakistan into Iron claw form both East and West. The tangible nature of these disputes makes them non-compromise and in-flexible issue for all involving parties. Major reason of the failed foreign policy of Pakistan is its weak institutional structure which remained the laboratory of experiments for every past ruler. The institutions were being used for personal will and interest. No clear definition for their framework being designed for them which ultimately results in overlapping and clashes among the state institutions. This clash in the machinery of state leads to uncertainty within the state as well as about its image at global level. This uncertain image leaves its impact on the political and economic front. Pakistan despite being a rich agricultural country has left way behind then its counter parts. The balance of trade is in critically negative situation. New agreements and trade deals are less and the present are not sufficient to meet the economic needs and interests of Pakistan.

References

1. Ahmad, D. I. (2006). Siachen: A by-Product of the Kashmir Dispute and a Catalyst for it sResolution. *Pakistan Journal of History and Culture, 27*(2).
2. AIAS.`(2007). *The Durand Line: History, Consequeunces and Future*, Istanbul.
3. Asia Report. (2012). Pakistan Relations with India: Beyond Kashmir? Washington DC: International Crisis Group
4. Alesina, A. (1997). *T.* Paper presented at the World Bank Conference on Development, New York.
5. Amir Ramin, N. S. (2013, January 2). Blood Line: Afghanistan and Pakistan's Unspoken Border Dispute. Retrieved from http://www.worldpolicy.org/blog/2013/01/02/blood-line-afghanistan-and-pakistan%E2%80%99s-unspoken-border-dispute
6. Baig, T. (2010). The Situation in the Afghanistan-Pakistan Border Regions. New York: UN Security Council.
7. Bajoria, J. (2008). Pakistan's Institutinos and Civil Society. Washington DC: Council for Foreign Relations.
8. Bajoria, J. (2009). THe Troubled Afghan-Pakistani Border. Washington DC: Council on Foreign Relations
9. Burke, S.M. Ziring, Lawrence (1990). *Pakistan's Foreign Policy: An Historical Analysis.* Karachi: Oxford University Press
10. Cohen, S. P. (2011). *The Future of Pakistan*. Washington DC: The Brookings Institution.
11. Fair, C. C. (2009). Time for Sober Realism: Renogitiating US Relations with Pakistan. *The Washington Quartely, 32*(2).
12. FPCCI. (2009). Iran. Islamabad.
13. Hasnat, S. F. (2011). Pakistan-US Relations n Slippery Grounds: An Account of Trust and its Deficit. *Pakistan Vision, 12*(1).
14. Hussain, I. (1999). Insitutes of Restraint: The Missing Element in Pakistan's Governance. *The Pakistan Development Review, 38*(4).
15. Hussain, I. (2012). Propspects, Challenges and Risks for Increasing Trade Between India-Pakistan Trade. Karachi: IBA.
16. Hussain, S. R. (2009). Resolving the Kashmir Dispute: Blending Realism with Justice. *The Pakistan Development Review, 48*(4).
17. Hussain, Z. (2011). Sources of Tension in Afghanistan and Pakistan: A Regional Perspective. Barcelona: CIDOB.
18. Husain, Safdar. Latif, Ijaz (2012). Issues and Challenges in Pakistan and Afghanistan Relations after 9/11. *South Asian Studies. 27*(1).
19. Hilali, A.Z (2013). Pakistan's Foreign Policy Objectives in the Post-September 11, 2001 Era. Islamabad: ISSI
20. Khan, H. (2012). Trust Deficit and Pakistan-Afghanistan Rocky Relations. Islamabad: PILDAT.
21. Khan, R. A. (2007). Sir Creek: Origin and Development of the Dispute Between Pakistan and India. *IPRI Journal, 7*(2).
22. Khan, S. (2012, August 24). Clash of Insitutions *The Express Tribune*.
23. Malik, M. I. (2003). Intitutional Failure in Pakistan. Islamabad: PIDE.

24. Mehmood, K. (2010). Pakistan-India Relations: A Pakistani Narrative. Islamabad: PILDAT.
25. Meyerle, J. (2013, November 29). Afghanistan-Pakistan Border: Back to Politics as Usual?, *The Diplomate*.
26. Naved Hamdi, S. H. (2012). The Opportunities and Pitfalls of Pakistan's Trade with China and Other Neighbors. *The Lahore Journal of Economics, 17*.
27. Rahman, M. (1996). *Divided Kashmir: Old Problems, New oppurtunities for India, Pakistan and the Kashmiri People*. Boulder: Lynne Rienner.
28. Report CRSS (2011). Pakistan Security Challenges. Islamabad: CRSS
29. Rizvi, H. A. (2014, September 10). New Oppurtunities in Pakistan-India Trade?, *PIQUE*.
30. Rizvi, Hassan Askari (2004). Pakistan's Foreign Policy: An Over View 1947-2004. Islamabad: PILDAT
31. Rizvi, Hassan Askari (2010). Democracy in Pakistan. New Delhi: CSDS
32. Riedel, Bruce (2012). *Deadly Embrace: Pakistan, America and Future of Global Jihad*. New York: Brookings Institutes
33. Salim Raihan, P. D. (2013). India-Pakistan Economic Cooperation: Implications for Regional Integration in South Asia. *Common Wealth Secretariat*.
34. Khan, S. A. (2007). Pakistan's Policy Towards Cenral Asia: An Exclusive Since 1991. *Research Journal of Area Study Centre*.
35. *The Tribune*. . (2014, May 11).

CHAPTER NO. 5

INTERNAL AND EXTERNAL CHALLENGES TO PAKISTAN'S NATIONAL SECURITY-RESTRAINTS WITHIN SOCIETY AND SYSTEM FOR THE DEVELOPMENT OF NATIONAL CONSENSUS AND SECURITY

Introduction:

"The weak and defenseless, in this imperfect world, invite aggression from others"

Quaid-è-Azam Muhammad Ali Jinnah

The state of Pakistan is still in its initial stages of political development. Academia has long been mourning over the lack of national security and consensus building to reach a consolidated national resolution, having firm political support and roadmap, to cope with internal and external challenges of the state. The state witnessed the creation in haste by the British Government to divide it and depart from the Sub-Continent. Muslims, the subjugated community of India, lacked the economic and political stature to run the state as compared to the Hindus. The state of Pakistan inherited the gamut of challenges, both within and without, with its birth. Founding a political setup best suited to the social realities of Pakistan was a cumbersome task and sill is an unfinished agenda. The challenges, the state inherited with its inception, immersed the quality resources and occupied state's policy engagements. Post partition emerging challenges post partition augmented the state's ordeal.

The concept of security has transformed over time and has assumed a consolidated outlook requiring a wholesome approach by the state to deal with various internal and external issues. Earlier concept of state security through military means has evolved in to a security pattern encompassing internal threats' handling. The developed states follow a pattern of making policies on national security at a level of interaction among different spheres of state policy and decision making. The National Security Councils of various developed states are glaring examples and the precedent was certainly set by United States. The above stated body is headed by the President and assisted by the heads of various concerned portfolios. The motive of this interactive assembly is to align the exterior and interior encounters of the state with the objectives of its existence, survival and security. All the policy options of foreign relations and inner issues are measured to give it a wholesome consideration ahead of giving out concerned decisions. This is the mechanism; the matured democracies have evolved, to ensure the state security. It guarantees the consideration of disparate views and the cleavages of system and society. However, the mechanism developed over decades when the political institutions attained maturity, which is not the case in developing democracies.

Pakistan is a developing state having immature democracy. The institutions are still evolving. Newly created state of Pakistan lacked the major state infrastructure which India enjoyed after division. Global and regional political landscape posed serious situation in haste and Pakistan had to adjust itself along with internal strife and divisions. Intellectual discourse has been emphasizing the failures of system to cope with such trials. This study is aimed at highlighting the restraints of society and system which hindered the national consensus on the issues of national security. This research study will revolve around some basic questions. How the inherent social divisions restrained the national consensus? How the evolving system was later affected by the internal developments? What are the reasons limiting the contours of Pakistan's foreign and defense policies augmenting external challenges?

Internal Challenges:

The internal challenges to Pakistan's national security highlighted in the previous chapter are divided as

- Internal disharmony
- Social challenges
- Economic Challenges
- Political Challenges

1. Internal disharmony:

Pakistani society is abounding with disharmony. There are various inherent factors widening these cleavages. The society was orchestrated with diversified groups united in one state. The religion was the uniting factor. But later the weak institutional framework of society denied the strengthening of unification. Rather, the gulf widened. Following three causes exacerbated the social stratification.

Ethnicity:

Pakistani society is the cluster of disparate ethnic groups. They were united to create Pakistan on the basis of religious ideology but the lingual, racial and tribal affinities dominated immediately after its birth. In 1948, it emerged with Bengali movement on the basis of language. The creation of Pakistan had lessened the powers of landlords enjoying autonomy and influence for centuries. They supported emergence of ethnic cleavages.

The demographic diversity is common in every society but it is enhanced and exacerbated by the social differences. In Pakistan, the economic disparities considerably increased this gap. A major portion of Pakistan's history went to the era of martial law. The dictatorial rule increased the disparities and broadened social gaps. The multiethnic essence of Pakistani society is shattering owing to the dominance of Punjab province. In Punjab also the dominant part is played by Northern Punjab. The quota system, which was formulated to give participation to all groups, is criticized over giving preference to Punjab

The ethnic communities minutely highlight the inherent weaknesses and shortcomings of a state structure in provision of social justice. Ethnic groups are more concerned about their security and rights. Pakistan presents a pertinent example of this phenomenon. In Punjab the caste system is so strong that it still dominates the politics. In Khyber Pakhtunkhwa and Baluchistan tribal affinities play major role in politics. Since partition, the influx of migrants from India and their increased role in economy and politics underlines the volatile political reality. The *Muhajir* and *Sindhi* dispute is at the heart of politics in Sindh. It is most diverse province in terms of ethnicity. This diversity was majorly carved by the influx of migrants earlier in 1947 and then in 1971. The later addition of Punjabis, Pashtuns and Balochis was due to the attraction of Karachi as economic hub. Sindhi population developed a sense of insecurity due to this rapid and continuous influx of 'outsiders'. The migrants still preserve their identity as 'Muhajirs'. Balochis had developed this sense of insecurity at very initial stage and the military operation by Bhutto augmented this manifolds (Breaking curfew-ref). Pakistan's creation resulted in a blow to 'Pakhtoonistan' dream of Afghanistan. The population of NWFP was majorly in favor of Muslim League but the Congress Ministry of Khan Sahib opposed this.

Punjabi population is around hundred million comprising of around 55.6 percent of total population. This province dominated the population since inception of Pakistan. It was always complained by other provinces that it dominated the politics, economy, bureaucracy and military. The reason was the huge population developed enough under the British Raj. The Punjabis were predominantly recruited in British military and were ahead of Pashtuns. They were educated and joined Indian Civil Service. They were sufficiently politically aware and were in position to control the politics of newly created state. Ethnicity soon assumed leading position in politics of Pakistan. Over the decades, insecurities of aggrieved groups increased. This issue has become political survival of politics of various parties is dependent over it. It further escalated by lack of serious efforts to mitigate it through consociationalism which may help to allay the fears and concerns. Rather, this divergence was always used as a political weapon and the elections of 2013 present clear manifestation of ethnic politics broadly on the basis of castes and provinces.

Sectarianism:

Pakistan is under threat of sectarian division undermining the national security. The clear division in the society surfaced plausibly in anti-Ahmadiya riots in 1953. Although the state suppressed it then by imposing first martial law but the acceptance of the movement by society and emergence of Islamic clergy as powerful political force initiated transformation of society on sectarian grounds. During the Zia era this was further increased through his Islamization policy clearly depicting hardline approach. The current wave of sectarianism emerged after the alignment to war on Terror. The sectarian attacks in Pakistan increased by 53 percent in 2012 as compared to 2011. The threat to Pakistan's national security is grave. The rationale of these militant groups, killing other religious groups, is acceptable to the masses and appeal to their religious beliefs developed majorly by clergy. The most clear division came forward is between Shia and Sunni. However, the seeds were sown in the anti-Ahmadiya movement in which Shias also favored the persecution of Ahmadi's. The most striking factor in division came during the Zia era by the imposition of Zakat as compulsory tax when Shias denied paying it.

The period before the martial law of General Zia the sectarian divisions and rifts were not that deep. Sectarian rifts and feuds became worst in early 90s in the aftermath of Afghan Jihad. Deobandi doctrine provided the soil to fight this war. Whole society was transformed eventually. State's support for a particular sect to achieve results in Afghan war negatively affected heterogeneous society of Pakistan. The Afghan war was actually proxy war for many actors. Saudi Arabian and Gulf States' support for hardliner Deobandi groups in Pakistan instigated Iranian response which already was in search of opportunity to export and expand its revolution. The rift between Saudi Arabia and Iran resulted in funding of sectarian groups in Pakistan. Its outcome was in bloodshed in Pakistan and current severe sectarian division. Pakistan had various religious minorities and sectarian communities. Different Muslim sects had been part of this society. Before partition these religious cleavages were overshadowed by the objective of independence movement. Later, the minority groups resorted to find sufficient space for their survival and security. Grey areas of security policies and state's ambiguous attitude towards certain grave security issues have led to current impasse.

Multiform Education System:

Education of the society on a particular pattern serves the national interest of the state. The state is responsible to provide education to its masses. It is the constitutional responsibility and an underlined demand of state-society social contract. When the state falters, the vacuum is filled by non-state actors. The varied pattern of education system challenges the social fabric. Multi-layered society further divides in to various fragments with varied mindsets. One education system was which Pakistan inherited in British legacy. British education system was a peculiar design to meet the requirements of a colonial power and it was ardently secular in contrast to the local education systems of Muslims and Hindus in India. Afterwards, the public sector's educational institutions deteriorated due to its incapability to meet the requirements of rapidly increasing population within given budget.

The budget allocation for education in 2011-12 was 533 million rupees rather it only increased to 679 million rupees in 2012-2013. Additionally, the social patterns of different areas of Pakistan require the child's education in religious seminaries instead of schools. The state has no control over the religious seminaries' curriculum or administration. Even the board to overlook these seminaries is no single rather there are five Wafaq-ul-Madaris belonging to different sects. The wave of Islamization during Zia's era created the environment of encouragement of this divided educational pattern. The religious seminaries further divided owing to the financial support of external actors. This situation intensified sectarianism in already meagre society.

Private educational institutions entirely follow their own designed curriculum. Around 26% children attend private schools. The syllabus taught in these schools plausibly differentiates the religious education from modern western education. The emphasis is put on western education. The books taught are published by Oxford University Press. Sharp criticism is drawn by the books of Pakistan Studies and Islamiat written by foreign authors taught in these schools. The content of these books clearly contradicts the radical and profound religious and cultural values of society developed since partition. Various private sector educational institutions are established and aided by USAID (United States Agency for International development). This puts a curtain between two spheres of society through educational inculcation. Education plays a vital role in directing a society within a targeted direction of national security and national goal. After the passage of eighteenth amendment, the portfolio of education falls within the provincial subjects list. The ethnic and regional diversity rather the politics being run on the basis of ethnic and regional issues pose a great danger as educational pattern will be further divided.

Social Challenges:

Social Transformation: The society of Pakistan is on course to majorly transform socially. The social structure changed to the violent society due to various reasons. Pakistan inherited various problems with partition. Due to permanent imbalance in traditional armed forces structure Pakistan's security threat increased which was to be addressed through more emphasis on militarization. From start of the state the fractured institutions constrained state to address the social transformations. Religious parties also got a chance to enhance their influence in society and exploit the religious sentiment of people, religion being the organizing principle of Pakistan. The religious wave further increased by the penetration of Saudi and Iranian interest and their management through a proxy conflict in Pakistani society. These factors gave air to extremism and violence in society.

The course of social transformation initiated with the emergence of religious politics in early 1950s. The religious fervor among the population which was fresh owing to independence movement was turned towards sectarianism. The first beneficiary was Maulana Maudoodi's Jamaat-e-Islami which was not part of Jinnah's Pakistan movement. Later, with the Afghan Jihad and Zia's Islamization opened the doors for social trends towards extremism. Interests of the external powers exploited the situation and tendency of violence augmented with sectarian drifts. The security conditions in Afghanistan after the forces of United States left the region had a direct impact on Pakistan's society. The 9/11 attack on United States and Pakistan's role as front line state erupted the fomented terrorist tendencies. The current wave of extremism, terrorism, insurgency and sectarianism are the eventuality of the decades of myopic policies.

Social Securitization: The Civil Armed Forces are responsible for the security of the state. Pakistan's police force is not equipped fully to perform this role effectively. The Police structure is the legacy of British colonial rule. The underlined feature of that structure was dealing with the 'masses as subjects'. The modern order of the day requires a detailed framing and restructuring of the Police Department. The lack of adequate equipment and training are the biggest hurdles to meet the challenge of social securitization. The Police Ordinance, 2002, was an effort to change the structure of police department and make it more efficient and autonomous. Yet, the training of police and resources are still a challenge to deal with.

Societal shift towards insecurity has certainly its roots in various external upheavals, in particular within the region and in general globally. But the responsibility to alleviate the impacts of such impulses was on the civil forces in charge of maintaining peace in society. The societal inclination towards extremism and violence of various forms delineate the bleak picture of an institutional impasse which remained short of saving the society from negative implications of external policies. Rather the system was not developed enough to cope with the realities of rapid transformation. Legacy of colonial rule continued to treat the masses as subjects which kept on depleting the state-society relationship. When the state resorted to side by the external goals at the cost of internal unrest, the society started slipping into 'black hole' of militancy. Perhaps the cost could have not been reckoned at that stage. But later too, it was always ignored. The institutional reforms in police were and are most pressing requirement to keep peace maintained in society.

Lack of Nation Building: The heterogeneous society like Pakistan needs a comprehensive policy framework to keep its population closely knitted and above the ethnic biases. Like many other issues of immediate importance nation building was never considered to be a national issue. The institutions never pondered to give this a shape through education of nation on uniform basis and engagement of people on a unified national goal. Growth of society in discrete manner and pursuit of diversified goals loosened the grip of state over society. Pakistan Movement collected people on a single platform for sometimes but after independence the politics assumed its earlier shape and ethnic and regional politics divided nation.

Economic Challenges:

The meagre economy of Pakistan is the greatest challenge hindering the way of progress and meeting all other challenges. The gamut of challenges requires huge effort with substantial resources to resolve all such problems. Pakistan was burdened with greatest influx of refugees in history immediately after partition. . It was augmented with the stoppage of its due share by India in the treasury of India before partition. Thus, it started its journey with thin pocket but having heavy burdens. Immediately after this, Pakistan needed to enter in to security pacts and started seeking for international aid and loans.

Pakistan is victim of dilemma of economy as the major portion of the resources goes for non-development funds. The foreign exchange is also used majorly for non-capital expenditures. By the end of June 2013, Pakistan's loan had reached around PRs. 14.5 trillion to PRs. 15 trillion. Mismanagement of resources and lack of bad governance created huge financial problems for Pakistan. The history of Pakistan's relations with IMF goes back to 1950 when it became its member and first loan it negotiated with IMF in 1958. This was standby agreement of USD 25 million which was later cancelled due to political reasons. The dependence has continued till date and the amounts of loans have soared to billions of dollars. Pakistan also faced a huge problem of allocation of resources to its provinces. The division among provinces always created bitterness in relations of provinces with center. Under the National Finance Commission award the allocation of resources has been sought out to some extent but the grievances of smaller provinces are deep rooted. Population was used as a measure of distribution of financial resources under NFC award due to which major portion goes to Punjab. The already soared relations between provinces are further deteriorated by this.

Political Challenges:

The state of Pakistan was in fight with the challenges which expanded the role and influence of its Armed Forces. This led to the intervention of armed forces in Pakistan's politics too. The lesser democratization of Pakistan is also due to the lose state-society connection. The political stability gives birth to nation building. The instability depends on the lack of adaptability by state and society. The change in one affects the other. The identity crisis emerging in the wake of political transformation has to be dealt with by the state. It is basically the lack of association with the state. The framework to plausibly define the role of institutions is also blurred.

Political Culture of Pakistan: Pakistan was created out of a unique part of world and a distinguished region of South Asia, the sub-continent. This region had a history of millennia and a culture of which it was always proud of. The castes and biradri system of this portion of globe had a unique outlook. Pakistan inherited this political legacy and eventually similar culture evolved here rather immediately after partition the politics of Pakistan took that course. The political culture of a state emerged out of the political history of that and political behaviors of its masses. Pakistan passed through same phenomenon. The democracy based on pure democratic values could not flourish in this society owing to its legacy of political culture which is the outcome of racial and family politics. Muslims had autocratic ruling system in sub-continent led by various Afghan, Turk and Arab dynasties. This in particular divided the Muslim population in to castes.

Pakistan inherited the parliamentary form of government as it inherited various traditions of British legacy. But the political culture did not have compatibility with this form of democracy. Theoretically there always remained consensus upon this form of government. The constitutions of 1956 and 1973 retained parliamentary democracy. British Raj endeavored to give best form of political institutions to India best suited to its communal differences. But the political realities of India were much different. The leaders of Muslim League could not conceive any other system apart from that model given by British Raj. The later undemocratic developments of in the political history of Pakistan were the eventuality of non-compatibility of political culture and political system. The onus of developing political institutions on the lines of democratic values was on the political leadership. The leadership could not deliver and lacked the initiative to legislate for the development of such political norms on which political institutions could flourish. The spaces in the political system gave way to military rules and long military interventions in politics.

External Challenges:

All the external challenges to Pakistan emerge from the fact of incapacity of the institutions to formulate the foreign policy and defense policy addressing the national requirements in totality. Immediately after the independence, Pakistan had to enter in to Capitalist bloc clearly defining its preference. It was the outcome of need of recognition and resources. This preference put immense pressures and constraints on Pakistan's foreign policy in coming decades. During upcoming decades Pakistan endeavored hard to look for a diversified policy but the global and regional scenarios demanded abrupt shifts in its policy. Thus, the immature and underdeveloped institutional framework could not deal with such bold realities and challenges affectively. The inherent threat to Pakistan came from India. This was the prominent proposition of Pakistan's foreign and defense policy. The defense of the state was to be secured in this perspective most. This led to a scattered and myopic policies dealing with short term and immediate threats and denied a chance to lay down a detailed policy framework with consolidated effort and long-term approach.

Pakistan is located in a region which always remained relevant to the global politics. It was created at a time when cold war had initiated. Its unique location at the edge of South Asia, separating it from Central Asia and a connecting route to South East Asia, earned it a status of an important state. India proclaimed neutrality while the world was clearly divided in to two blocs. Pakistan followed same course initially but later succumbed to the internal restraints and challenges. Pakistan inherited a volatile environment. Security dilemma persisted always with India which led Pakistan to exacerbate its search of security alliances.

Foreign policy impulses:

The foundation stone of Pakistan's foreign policy was placed by Quaid-e-Azam. His rhetoric manifested his profound allegiance to democratic principles. Yet his discourse depicted strong commitment to the religious principles as the political elite of Muslims in India overwhelmingly used religion to maneuver masses. Iqbal and Jinnah had a shared dream to establish a modern democratic state in Pakistan shunning the theocratic trends of earlier Islamic empires. It has always been blamed by an ilk of Pakistani intelligentsia and academicians that Islamic ideology was fiercely used by the Muslim leaders to ignite popular sentiment leading to founding of state. The 'identity crisis' prevalent in society currently emerged from enormous pressures arising out of transforming global and regional political dynamics. Foremost requirement of the newly created state was to find a way out of its isolation and sought a mode to strengthen its security. The institutional impasse deeply affected Pakistan's foreign policy formulation too. Pakistan's foreign policy has always succumbed to reactive approach and myopic concentration and planning of interests.

Pakistan's imminent and foremost security threat emerged from India and it soon was confirmed by latter's naked acts of aggression against the very survival of it. Pakistan's first priority remained to make its security strong on eastern border. Sir Zafrulla Khan, first foreign minister of Pakistan, stated that that bilateral relation could be immensely positive and decisive in world affairs but both had been pushed in to conflicting positions which could disrupt global peace. With partition, Pakistan inherited gamut of problems while the resources and civil and military administrations were in the process of division. While the newly created state had to raise up its institutions matching the very ideological basis of its foundation, it was faced with unimaginable partition by Boundary Commission headed by Cyril Radcliff. Quaid-e-Azam averred the award 'unjust, incomprehensible and even perverse'.

This was followed by influx of refugees whose rehabilitation was cumbersome with meagre resources. The unjust border division gave birth to the water dispute lingering on till date rather intensified manifolds. Another problem which further made Pakistan's security fragile was the closure of Supreme Commander's Headquarters before its mandated time which was 1st April, 1948. The then Defense Minister of India assured just division of military equipment and advocated its closure but later withheld Pakistan's share inflicting huge loss. It was felt by Pakistan and later proved by Indian Aggression in Kashmir and its military accession that the attack was on very foundations. Later, insecurity on part of Pakistan substantiated from India's rapid military development and show of military strength in the war of 1965. Soon after this Indian involvement in Bangladesh and division of Pakistan in 1971, gave birth to new sense of anxiety which led to nuclear race. Various major crises also erupted with India continuing the hostile trend between the two. India's foreign policy also depicted sharply its designs to counter Pakistan. Most plausible example is of India's support of Afghanistan's aggressive and negative diplomacy towards Pakistan. Throughout history India supported Afghanistan against Pakistan followed by its doctrine 'neighbor is your enemy but neighbor's neighbor is your friend'. Indian policy towards

Afghanistan was narrowly in pursuit of its opposition to Pakistan. Pakistan's Afghan policy remained reactive owing to Indian involvement there. After US invasion succeeding 9/11, doors once again opened for India which had been closed due to Taliban rule. India's massive investment of $1.2 billion in just infrastructure depicts its aspirations to encircle Pakistan and thin down its opportunities. India remained a central security threat to Pakistan and compelled it to take reactive short term decisions.

India's defense policy focus also remains Pakistan. For last three decades India paid considerable attention to its military up gradation with specific reference to Pakistan. Pakistan was put top of the list concern in Indian Ministry of Defense's reports since late 1980s. The outlook of Indian policy might have changed but the core issues of dispute between India and Pakistan remain there rather the intensity and depth of these conflicts have considerably increased. Kashmir is on top of this list of bilateral disputes. The internal image of this issue has been reduced to bilateral engagement. But the subjective dimension of the issue for both states is grave. The water dispute lingers on till date and has assumed severity till today.

During the initial years Pakistan followed the policy of being non-aligned as India adopted but it could not retain the policy owing to hostile regional environment. Pakistan and Afghanistan inherited soared relations with British colonial rule who separated Pashtun land. Afghanistan's claim over Pashtun area was the basic frictional point between the two. Afghanistan was the only state which opposed Pakistan's admission into United Nations and conditioned it to right of self-determination to people of NWFP India's involvement in this led to Pakistan's deep concern over Afghanistan. Later all policy decisions on part of Pakistan can be associated to its deep insecurities in Afghanistan. The rapid decision to move in to Afghan dispute in the wake of Soviet invasion was a move to enhance its role and influence on Afghan soil to mitigate the severity of relations. Afghanistan has not accepted Duran Line yet. Pakistan went beyond its capacity to give shelter to Afghan refugees. But the rocky relationship continued rather augmented after 9/11. Pakistan's association with Taliban has always been condemned by other Afghan groups, Northern Alliance in particular. Afghan issue has most deep and far reaching impact on Pakistan's security and a close nexus with internal security and stability issues.

Pakistan found an opportunity in Soviet attack on Afghanistan to defend itself on western border. United States' of giving birth to Jihad in Afghanistan suited Pakistan and both entered in to another era of close partnership. Zia regime endeavored to find solution to various problems through this war. Pakistan was out of relations with United States which was important to rejuvenate and upgrade military drastically. Zia's military coup was in the process of initial years and it needed legitimacy which this war could come up with. And longstanding history of hostile relations with Afghanistan gave a lesson to assuage Afghan hostility by increasing influence on that soil which Pakistan achieved by deepening relations there under cover of Jihad against Communism. The influence continued with Taliban but at the cost of agony and enmity of other ethnic groups of Afghanistan joined as Northern Alliance. The ethnic and religious grouping in Afghanistan suited United States which it used for its benefits against Communist Russia This alliance having considerable amount of Shia population enjoying support of Iran augmented the bilateral connection of security concerns of Pakistan within and without. Saudi Arabia played key role by supporting Deobandi Mujahideen factions too. The concatenation had substantiated the argument of sectarianism which has now become a monster for Pakistan's security.

Pakistan had presented itself as an important actor for United States' interests and the partnership in phases renewed with 9/11. Pakistan had brought refugees in to its territory which was succeeded by a violent culture. Pashtun affinity of Afghans with people of tribal belt in Pakistan created more space for them in Pakistani society. Pakistan's alliance with United States in War on terror earned it the status of 'most allied ally' but the power spectrum it had created with Afghan Jihad shattered. So many Deobandi factions defected in reaction and insurgency against the state initiated. Although state had launched operations in tribal belt to melt down the intensity of this militancy but is still in state of war. The phenomenon of Saudi-Iran rivalry is clearly being played out on Pakistani soil which has linked to Pakistan's internal security concerns through policy on Afghanistan. The Middle Eastern Shia-Sunni rivalry has become an open sore for Pakistani society in the form of sectarianism which is at its most dangerous ebb.

Pakistan inhabits considerable percentage of Shia population and there was no such hostile environment as it is today. During the initial years, Pakistan showed extra ordinary effort to join hands with Muslim World which was responded coldly. However, the relations with Iran started on exemplary pace and deep understanding of each other. Iran being the first state to recognize Pakistan became a top priority in Pakistan's foreign policy framework. Pakistan's tilt towards United States was also a convergence of interests for both states as Iran and United States had deep rooted relations during the era of Pehalvi Dynasty. The Iranian revolution brought a sudden change in the regional matrix and the succeeding Shia theocracy in Iran narrowed the options. The revolution though was not threat to Pakistan yet the immediately following Afghan Jihad and Pakistan's policy options in Afghanistan created a rift in bilateral relations. Iran's effort to export its revolution augmented the trust deficit. There are so many spheres where both states can have enhanced cooperation in particular in energy, regional security and trade. Iran's relations with India stirred more bitterness in the relations.

Pakistan-US relations are the corner stone of Pakistan's foreign policy owing to its dependence on United States. Pakistan's first Prime Minister Liaqat Ali Khan's first visit to United States initiated the relations. Pakistan's vulnerable economic conditions after partition compelled it to look towards United States which was viable option as compared to relations with Russia. Pakistan entered in to SEATO and CENTO pacts in search of security from India but could not realize that those defense pacts strictly triggered response in case of Soviet aggression. Pakistan was an important state in South Asia to keep regional balance. Afghan Jihad made Pakistan critically important. Since inception Pakistan needed military aid which United States provided considerably to meet Pakistan's security requirements. Pakistan deemed it important to be relevant in global politics as isolation could negatively affect its security designs. It witnessed this insulation and seclusion during phases of cold relations with United States. So, Pakistan had limited or rather no choice at all after 9/11 except to join the US camp once again.

China was the state with which Pakistan's relations started on a warm note but later Pakistan's strange role on the event of China's membership to United Nations created mistrust. Later, Pakistan responded in purely pro United States' manner in issues of Tibet and Taiwan. But later the relations revived again. Basically the rejuvenation of bilateral relations was based on needs and regional security. On India the concerns of both states converged. Pakistan is a strategic need of China and Pakistan has this card in a thin pocket of choices. Pakistan found it highly feasible to keep relations warm with China. For Pakistan's security requirements China is a time tested partner which tries to exploit Pakistan's various resources to keep its trade and manufacturing continuous.

Enhanced role of military in external affairs: Democracy remained meagre in Pakistan due to various internal vulnerabilities. Major period of Pakistan's total age went to military rules. The quest of security since the early period compelled Pakistan to rely on its army for the settlement of external affairs. Two wars with India during the early phase and dismemberment of one of its part (East Pakistan) gave the bitterest lessons. Pakistan is situated in the geography of tumultuous politics and hostile environment. The emphasis became military development which further empowered already organized and most disciplined institution which inherited the legacy of British Royal Army. Later, army's liaison with United States and engagement in various security spheres of bilateral interest strengthened it.

Conclusion:

The list of challenges to Pakistan's national security is long and cumbersome to deal with simultaneously. The commonality in the external and internal challenges is the incapability of the state institutions to curb these threats. Pakistan is a developing nation which inherited various problems and dents with its independence. The constraints and limitations out of the control of state further deteriorated the security and increased the national security threats to many folds.

The national security of Pakistan has always been an unfinished agenda for the state. The nature of challenges to Pakistan's security, both internal and external challenges, remained so intricate which its institutions, under evolution and development, could not address in totality. Pakistan came in to existence with a gamut of challenges in inheritance. The immediate threat emerged from the border insecurity from India and Afghanistan. Pakistan witnessed the Indian aggression immediately after partition in 1948, through Indian quest and endeavors to accede Kashmir militarily. On the western border Afghanistan remained a threat to Pakistan's external security front. With the passing developments of global and regional political ambiance, Pakistan's external challenges augmented.

The institutions of Pakistan are primarily the legacy of British colonial rule. The internal environment of the state developed under the shadows of external threats, the ethnic diversification further deepened by extremist and sectarian trends and intensified by multiform education systems and curriculum. The economic degradation owes its birth to constraints of the meagre economy depleting by social and political factors. The society of Pakistan transformed negatively due to absence of suitable social security pattern. The political system of the state is in evolution still. The domain of functioning of political institutions remains blurred due to the ongoing process of development and the pattern of political culture of this society. This study will assess the institutional limitations which kept it incapacitate to deal with such challenges.

CHALLENGES	WHAT	WHY
Internal Challenges	Internal Disharmony	• Ethnicity • Sectarianism • Multi-form Educational System
	Social Challenges	• Social Transformation • Lack of Nation Building • Social Securitization
	Political Challenges	• Political Culture • Non-compatibility of Political Institution & political culture
	Economic Challenges	• Economic Vulnerabilities since partition • Reliance on aids & loans • Issue of distribution of financial resources among provinces
External Challenges	Foreign Policy Compulsions	• Narrow Options • Security threats from India • Weak Institutions

	Emphasis on Defense Policy	• Continued Security threats from India • Three wars & various crisis from India • Nuclearzation of region
	Enhanced Role of Military in Foreign Policy	• Search of Security • Emphasis on military development • Race of Nuclearization
Internal & External	• Lack of Accountability • Lack of Responsibility • Sharing and Institutional Harmony	

References:

1. Ahmed, M. (2006). Political Culture in Sub-Continent. *Al-Siyasia: A Journal of Politics, Society and Culture*, 36-49.
2. Ahmed, M., & Naseem, F. (2011). Social system influences Political System; A comparative study of sub-continent. *Berkeley Journal of Social Studies*, 1-11.
3. Alam, S. (2004). Political and Strategic Dimensions. *Strategic Analysis*, 526-546.
4. Ali, A. (2010). The NAtional finanace Commision Award: A Way Forward. *Reflections*, 1-5.
5. Bajoria, J. (2009, October 7th). *Pakistan's Education System and Links to Extremism.* Retrieved from Council on Foreign Relations: http://www.cfr.org/pakistan/pakistans-education-system-links-extremism/p20364
6. Burke, S. M., & Ziring, L. (1973, 1990). *Pakistan's Foreign Policy: An Historical Analysis.* Oxford, Karachi: Oxford University Press.
7. Burke, S. M., & Ziring, L. (1990). *Pakistan's Foreign Policy: An Historical Analysis.* Karachi: Oxford University Press.
8. Butt, U. (2010). Pakistan's Salvation: Islam or Western Inspired Secular-Liiberal Democracy? In U. Butt, & N. Elahi, *Pakistan's Quagmire: Security, Strategy and the Future of Islamic-Nuclear Nation* (pp. 9-37). London: Continuum.
9. Durrani, M. U., & Khan, A. (2010). Pakistan-Afghanistan Relations-Historic Mirror. *The Dialogue*, 25-65.
10. Finance, M. o. (2013). *Budget in Brief.* Islamabad: Ministry of finance, Government of Pakistan.
11. Hussain, M. (1993). Indian Power Projection in South Asia and the Regional States' Reactions. In H. Malik, *Dilemmas of National Security and Cooperation in India and Pakistan* (pp. 152-163). London: Macmillan Press Limited.
12. Hussain, S., & Latif, M. I. (2012). Issues and Challenges in Pakistan-Afghanistan Relations after 9/11. *South Asian Studies*, 89-99.
13. Iqbal, Z. (1995). Constraints to the Economic Growth of Pakistan: A Three-Gap Approach. *The Pakistan Development Review*, 1119-1133.
14. Javaid, U., & Mushtaq, I. (2014). Historical Perspectives of Pakistan USA Relations: Lessons for Pakistan. *South Asian Studies*, 291-304.
15. Kennedy, C. H. (1984). Politics of ethnic Preference in Pakistan. *Asian Survey*, 688-703.
16. Kennedy, C. H. (1991). The Politics of Ethnicity in Sindh. *Asian Survey*, 938-955.
17. Kiran, A. (2011). Conceptual Framework of Homeland Security: A case study of Pakistan. *NDU Journal*, 187-220.
18. Majeed, G. (2010). Ethnicity and Ethnic conflict in Pakistan. *Journal of Political Studies*, 51-63.
19. Memon, P. A., & Memon, K. S. (2011). Political Instability: A case of Pakistan. *Journal of Political Studies*, 31-43.
20. Mushtaq, M. (2009). Managing Ethnic Diversity and Federalism in Pakistan. *European Journal of Scientific Research*, 279-294.
21. PIPS. (2013). *Sectarian Violence.* Islamabad: Pakistan Institute of Peace Stidies.
22. PWD. (2014, August 23). *Population Welfare Department, Punjab.* Retrieved from Population Welfare Department, Punjab: http://www.pwd.punjab.gov.pk/

23. Ramzan, M., & Javed, U. (2013). Police Order 2002: A Critique. *Journal of Political Studies*, 141-160.
24. Rehman, I. (2012). Drowning Stability: The Perils of Naval Nuclearization and Brinkmanship in Indian Ocean. *Naval War College Review*, 64-95.
25. Review, E. (2013, October 2). IMF and PAkistan: The Past and the Present. *Economic Review*, p. 1.
26. Rizvi, H. A. (2012). *National Security Council- A Debate on institutions and processes for decision-making on security issues; Civil-Military Relations.* Islamabad: Pakistan institute of Legislative Development and Transparency.
27. Sattar, A. (2013). *Pakistan's Foreign Policy: 1947-2012*. Karachi: Oxford University Press.
28. Shafqat, S. (1998). Democracy in Pakistan: Value Change and Challenges of Institution Building. *The Pakistan Development Review*, 281-298.
29. Shafqat, S., & Haq, R. u. (2011). *Pakistan, Afghanistan and US Relations: Implications and Future Directions.* Lahore: Center for Public policy and Governance.
30. Shahzad, A. (2008). Ethnic Conflict in Pakistan: Historical Background. *Pakistan Vision*, 125-166.
31. Syed, A. (2013, January 1). Each Pakistani inherits Rs. 80,000 as loan burden. *Pakistan Observer*, p. 1.
32. Waseem, M. (2011). *Patterns of Conflict in Pakistan: Implications in Policy.* Washington DC: Saban Center-Brookings.
33. Weinbaum, M. G., & Harder, J. B. (2008). Pakistan's afghan policies and their consequences. *Contemporary South Asia*, 25-38.
34. Whittacker, A. G., Brown, S. A., Smith, F. C., & McKune, E. (2011). *The National security Policy Process; The National Security Council and Interagency System.* Washington DC: The White House.
35. Wirsing, R. G. (1993). The Kashmir Dispute: Prospects for Conflict Resolution. In H. Malik, *Dilammas of National Security and Cooperation in India and Pakistan* (pp. 164-206). London: Macmillan Press Limited.
36. Yousafzai, M. R. (2011). *Pakistan-Afghanistan Relations: A Pakistani Narrative.* Islamabad: Pakistan Institute of Legislative Development and Transparency.
37. Yusuf, H. (2012). *Sectarian Violence: Pakistan's Greatest security Threat.* Norwagian Peacebuilding Resource Center.
38. Zaman, M. q. (1998). Sectarianism in Pakistan: The Radicalization of Shi'i and Sunni identities. *Modern Asian Studies*, 689-716.

CHAPTER NO. - 6

NATIONAL SECURITY: ROLE OF MEDIA

In *Politics among Nations*, Hans Morgenthau defined national security as "the integrity of the national territory and its institutions" (Morgenthau, 1948). It is to preserve the sovereignty of the state and safeguarding its national interest from within and outside. It has a close relationship with what is called the fourth pillar of the state, 'the media' which is considered to be the watchdog and a reflection of the society.

The previous decade marked the advent of many private TV news channels in Pakistan. Transparent, free and independent media were the mythical words, Pakistanis uttered and celebrated but in its infancy, adolescence and now in teenage years, media in Pakistan has not played its role with maturity when it comes to the matters concerning national security and national interest. Why is that? It will be explored in this chapter.

This study is designed in the context of Pakistani society, its media and the national security and strives to highlight the weak areas Pakistani media needs to work upon to secure country's national interest and how the media should refrain from irresponsible reporting of events concerning the matters of national interest and national security. To watch the watchdog is the aim!

September 11 and the U.S Media

It is said that the media has the power to make or break societies and for the act of making and breaking, it may employ many tactics. September 11 is a splendid illustration of how U.S media strengthened and supported the policy of the Bush administration in the favor of war on terror. Douglas Kellner uses the term *'spectacles of terror'* for the coverage of world trade center attacks and how it was broadcast throughout the world to show how vulnerable America is to such terror attacks.

Broadcasting, on September 11, was an act of repeating and replicating the incident and to make eyes and ears hostages of what the Muslim terrorists of Al-Qaida unfolded for the U.S and its national security. It was shown as a division between the good and bad, just and unjust and a battle to wipe out the necessary evil from the society.

Many national security intellectuals were brought on screens claiming that it's a war between Islam and the West and through the media coverage Huntington's clash of civilizations was much more evident than ever before. The media framed the whole crisis within the context of Islam, of cultural conflicts, and of Western civilization threatened by the other (Abrahamian, 2003). Thus the need for eradicating the evil from society was created and fully supported by the U.S media.

Stephen D. Reese and Seth C. Lewis in their article *framing the war on terror: the internalization of policy in the US press state that* the bush administration introduced the label war on terror to its national security policy after the attacks of 9/11 which was fully backed by the US media. "The War on Terror became a socially shared organizing principle through its transmission via the US press" (Resse & Lewis, 2009, p.778). One may call it media manipulation, its support in favor of national security of the U.S or media framing to support the Bush doctrine of war on terror, it supports the U.S national interest in every possible way.

If one critically analyzes the U.S media coverage of the September 11 attacks following the war on terror, one phrase can summarize it and that would be *'identification of the enemy.'* Right after these attacks, CNN, Fox and many other U.S media channels started identifying their enemy and shaped the minds of millions in the favor of U.S national interest and later supported Bush administration's military intervention strategy against the enemy.

The international media, the biggest supporter of freedom of expression, tends to flow in the direction of national interest and sometimes censors the other side of the story. The war on Iraq was indirectly supported by the excessive projection of violence on television that increases the aggressive nature of the American culture, its society, and its politics (O'Hanlon, 2003). The direct support for the Iraq war and the war on terror has been done with the help of broadcast media forums such as CNN, Fox news and MSNBC and many others, she articulates, "News broadcasts disseminate biased information and present selective imagery, thereby constructing an inaccurate public reality and legitimizing violent political practices in attempts to rally public support for the War on Terror and a war on Iraq." This is where their national interest reigns supreme.

Pakistani Media Falls Short of Identifying the Enemy of the State

As a comparison with international media, in Pakistan, every Tom, Dick and Harry seems to know what the national interest is and they decide it for themselves (Manzar, 2014). There has always been a question in my mind that in a comparison with international media, have Pakistani media ever played its role in the identification of the enemy or the media itself is the enemy of the national interest?

As an effort to answer this question, media has the power to unite even in the worst circumstances and can bring the divided nations on a common platform with the power it possesses. But has our media done that? During the past decade and in its coverage of the war on terror, Pakistani media did not bring the nation on one stand. The nation is divided because the media is divided. The price for media affiliation with different political parties and its sources of funding has to be paid by the public. Media could have led us to unite on this war on terror and the media could have presented the facts to people rather selling emotional videos and talk shows to the public (Emmjay, 2008).

According to Bernard Cohen, it is the media which draws the political maps of the world (Cohen, 1963). With this claim, a great responsibility comes on the shoulders of media to portray the positive image of the country. During the previous decade of its coverage of the war on terror, all media did is to criticize the policy and to promote people's anger and emotion; why wouldn't they have done this, after all it was the selling product. They sensationalized every aspect of the war on terror to sell it more to the public and a little importance was given to being responsible (Emmjay, 2008).

"We (Pakistani media) have not been able to identify our enemy due to the certain interest of the certain groups (Rauf, personal communication, August 30, 2014)." Today media is not just a dutiful profession but the corporate culture dominates it. And in the pursuance and projection of interest of a few, it sometimes squeezes the national interest of the country. The one thing which is slaughtering the notion of a responsible media is the culture of sensationalism. What is the point of interviewing militants like Mullah Fazl-ullah and others? Statements by terrorist organizations get a highlighted place in our media. Ratings-hungry TV channels eventually brought catastrophe to Pakistan. Their popular anchors gave space and sympathy to murderers and terrorists, and broadcast every lie, rumor, and idiocy that could sell (Hoodbhoy, 2014). A widespread confusion has been created by Pakistani media. People are unaware of their enemy and what war they are fighting and why? A confused media is building up a confused nation and is contributing in creating a sense of restlessness in the society which is already doomed with it.

Media and national security in Pakistan

By looking at the national security strategies of the countries with the world's strongest media outlets, U.S. and U.K, one can see that these strategies do not include any directions or input for the news organizations. On the contrary, the newly made national internal security policy by the ministry of interior includes provision for media in its clause 42 under the heading contours of NISP and articulates "to prevent misuse of social, electronic and print media" (NISP, 2014).

This may raise a question in the minds of many that when countries like America and Britain don't have any portion in their national security strategies which guides media what to do then why Pakistan's national internal security policy has such a point which provides directions to media. Is media in Pakistan going off track? To answer this, a couple of factors need to be discussed.

A culture of commercialism, sensation and thrill

Gone are the days when media and journalism were considered dutiful professions. The corporate culture and business minded approach have commercialized it to the extent which is beyond imagination. Today, even a drop of tears is being commercialized in the media. Robert McChesney puts it in a quote, *"if you look at the history of broadcasting, what you find is the National Association of Broadcasters is a trade association whose mission is to protect the interests of the commercial broadcasters."*

Dr. Nazir Hussain in his research article *Role of Media in Internal Security of Pakistan* says that "media can never be independent: either it works under the government pressure imposed through laws or under the influence of financer, as it has become a moneymaking industry" (Hussain, 2008). Media, today, is more business oriented and in the pursuance of improving its ratings and to break the news at its earliest, it is becoming more prone to sensationalism and trill. Pakistani media is no exception to this. Gate keepers distort and filter the reality and serve the purpose of sponsors who fund them.

Scattered body parts, stains of blood and a wave of terror are what eyes watch and ears hear being reported on Pakistani channels after a bomb blast or suicide attack. Where happens this kind of reporting in the world? Certainly nowhere. The dramatic tone, body language and trembling voice of the reporter give it even more hype. Media in Pakistan follows if it bleeds, it leads principle. Reporting is an objective area of journalism but sensationalism has killed the true meaning of reporting. Pakistani channels are getting sensationalized to a greater extent to use such sad incidents as their selling points which is threatening to the internal and external security of the country.

Terminology and Language used by the media

Terminology and use of language by the media has a direct impact on the masses. As George Orwell in his 1946 essay with the title *politics and English language* states that, "if thought corrupts language, language can also corrupt thought" (Orwell, 1946). The language used by media persuades people in understanding meanings of events. It also affects their knowledge and interpretation of events. David Buckingham in a UNESCO published thesis of Thessaloniki with the title *questioning the Media: A Guide for Students* gives language used by media utter importance and highlights four main areas to consider regarding questioning the media which are:

- Production: who makes media?
- Language: how do media communicate meaning?
- Representation: how do media portray the world?
- Audience: how do people make sense of the media? (Buckingham, 2003)

BBC's editorial guidelines, to ensure objectivity, aim not to use the term terrorist without attribution. The case is quite the opposite in Pakistan. The language used in media reporting can cause unrest in society particularly on the issues regarding terrorism, sectarianism and ethnicity. Let's take the example of terrorism reporting by Pakistani Media. Regular TV users can measure the use of different terminologies used by Pakistani media. Different media channels have different terminologies to describe terrorists. Some call them militants; others call them fascist or 'shidat pasand'. In same kinds of incidents, some people are being called 'Halak' and others 'Shaheed or Janbahaq.' It is such a dangerous thing (Razi, personal communication, September 9, 2014). This causes confusion among people in the interpretations of events.

Irresponsible reporting of events

No one can deny that freedom of expression is the integral part of an open society but it should not be used as a cover to harm the national interest of the country. No open society would allow that. In the past, after the Bombay attacks we saw newsbreaks flashing on Pakistani TV channels claiming that Ajmal Kasab belongs to Pakistan when India was already accusing Pakistan for those attacks. Such a sensitive scenario demands to preserve the national interest of the country rather breaking the news at its earliest and fulfilling the claims of freedom of the press.

The point to ponder is that "*not all information is made to be revealed to public. Necessary Secrets: National Security, the Media, and the Rule of Law* are a book by Gabriel Schoenfeld *which presents a conflict between freedom of the press and secrecy of national security boundaries*". He examines the tensions betweenthe government's need to preserve certain secrets and the role of a free press.

In this book, headvocates that, "Along with the public's 'right to know,' constantly invoked by the press, there is also something rarely spoken about let alone defended: namely the public's right not to know" (Schoenfeld, 2011, p. 258). He calls this a threat to the national security of the U.S and criticizes journalists and press which consider themselves what he calls "the sovereign power, above the three branches, and free to violate their democratically enacted laws in pursuit of its mission" (Schoenfeld, 2011, p. 53). In this book, Schoenfeld has made quite adequate arguments. While reporting facts to the people, press and media should consider the limitations and responsibility they are restricted to and what can possibly harm the national interest of a country.

But a recent example of a particular media group presents a completely different picture which, after the attack on journalist Hamid Meer, burst into saying that it was done by the head of the Pakistan's military security agency. In addition, running the clips of the DGISI by that particular media group for hours gave chances to many previously zipped mouths to open up and to malign Pakistan's image internationally.

Media always promotes its notion of free access and reporting of information. This is not to be criticized but when issues such as national security are involved then media cannot always be allowed to report whatever it aspires and in whatever manner it wants. Dr. Nazir Hussain in a 2008 research report by South Asian Strategic Stability Institute with the title *The Role of Media in National Security: A Case Study of 1998 Nuclear Explosions by Pakistan* states that "When the two institutions–military and media–meet during a conflict, clashes are expected." (Hussain, 2008) It can be related to the military vs. media clash in the aftermath of attack on Hamid Meer. He further sheds light on this by saying that the media wants freedom, no censorship, total access and the capability to get to their audience quickly. The military on the other hand, wants control.

With reference to the above mentioned example, where on earth, the country's own media do harm this way to the national interest and security of the country? Rather, media helps elements of security agencies in different matters as Orya Maqbool Jan puts it, "secret agencies and the media have a long history of association and this relationship continues and will keep on continuing." He further adds, "The secret agencies, media and courts do not oppose one another in any of the democratic countries. That happens only when people feel that their newly found media power is too great. They start feeling that they have become king-makers and policy makers. They start feeling that they can make and break governments. These people should look around to see the ground realities and study history" (Jan, 2014).

Flawed television rating system

Rating hungry TV channels have brought things to such a devastating brink where they want to reach on top of the sky of TV ratings. It is their only parameter to judge the popularity of the show by crushing the parameters of responsibility, credibility, authenticity and accuracy. Every channel claims that it was them which broke the news at its earliest. Focus on being the first to report and not necessarily the most accurate or truthful has created credibility issues (Ali, 2012, p. 246).

What is a TV rating system and why channels are made about it? It is a system based on placing people's meters with television sets in homes to view and note the timing that how much a program has been watched. These meters are not many in number and they have been put in big cities i.e. Lahore, Karachi and Islamabad etc. From the available meters, a large number is being put in Karachi and there is, even not a single meter erected in Baluchistan.

The popularity of the programs is measured from this flawed system. How such a small fraction can determine what is popular or not? Moreover, the advertisers give ads to those channels and programs which have high ratings so it's like a never ending war among media houses to be ratings-hungry. "Our people now have a tool in their hand called 'the rating system' which is so pathetic. If the rating of a certain channel or program is high, they will watch those programs. How can you imagine news channels showing programs which have 'bhands' and 'maskharas' making fun of everything and it is so substandard or showing the crime dramas" (Rauf, personal communication, August 30, 2014). Channels do it just to add crispness and to make it juicy.

The chase of getting high ratings have made political talk shows not more than a blame game where the anchors let participants abuse one another in such a language which is shameful rather than carrying out a debate. "Anchors like Mubasshar Luqman, who are totally sensationalizing and exploiting the situation to their benefit" (Rauf, personal communication, August 30, 2014). Talk shows, incessantly, more often deal with petty bickering, screaming matches between opponents invited, also leading to colorful exchange of abuses rather than policy matters or a constructive discussion (Ali, 2012, p. 247).

No efforts by media to educate Pakistani people regarding national security

Under the agenda setting theory of communication, Bernard Cohen argues that, "media may not be successful much of the time in telling people what to think, but it is stunningly successful in telling its readers what to think about" (Cohen, 1963). This famous quote of Cohen highlights the agenda setting role of Media. "Media has the power to highlight important issues, events and debates as well as it can undermine by ignoring them" (Eijaz, 2012).

Framing is an important concept under the agenda setting theory of mass media. Entman defines it as, "framing essentially involves selection and salience. To frame is to select some aspects of a perceived reality and make them more salient in a communicating text, in such a way as to promote a particular problem definition, causal interpretation, moral evaluation, and/or treatment recommendation for the item described" (Entman, 1993, p. 52).. Media frames issues to support a particular mindset. It can either be the agenda of state or the sponsor.

In the context of Pakistani society, lack of framing or let it be called lack of conscious faming is one of the major reasons behind people's confused status regarding national security. Media tells people what to think about but not that much efficiently. The time media gives to the reporting of a particular issue makes that issue important for people. Sadly, Pakistani media has never created awareness regarding national security and regarding preserving and enhancing it.

PEMRA's Code of Conduct: How Pakistani Media is Violating this Code of Conduct?

Pakistan Electronic Media Regulatory Authority was created in 2002 to have an eagle eye on the activities of Pakistani media houses with its ordinance. Unfortunately, it has just become a license issuing authority because of the excessive violations by media channels of its code of conduct. PEMRA's 2012 content regulations point no. 6 under the heading *coverage of incidents of accidents, violence and crime*with sub point 3 states that, "scenes with excessive violence or suffering such as close-up shots of persons being brutally tortured and killed shall not be shown" (PEMRA content regulations, 2012, p. 4). It is violated by almost every channel which shows the scattered body parts in the coverage of a suicide attack.

Pakistani Media gives coverage to the statements of militant groups and organizations and provide them an avenue for the projection of their views. PEMRA's content regulations point no. 12 under the heading *general standards* with the sub point no 2 states that, "licensee shall ensure that all proscribed organizations, militant groups or individuals who are notified by the government as terrorists shall clearly be so identified and statements of such militant groups or individuals shall not be aired" (PEMRA content regulations, 2012, p. 9).

PEMRA's code of conduct for media broadcasters or cable TV operators under point 1 and sub point 'g' states that "no program shall be aired which contains aspersions against the Judiciary and integrity of the Armed Forces of Pakistan" (Code of conduct for media broadcasters and cable TV operators, 2012). This is what people of Pakistan experienced lately when a particular media group accused Pakistan's military secret agency's head for the attack on one of its journalists. It shows that how much our electronic news channels care about PEMRA and its code of conduct!

The tragedy with PEMRA is that it became an authority like there are other institutions where nepotism is the culture. (Razi, personal communication, September 9, 2014) The problem is that Pakistan has a media regulatory authority which has its regulatory ordinance but it is not followed by media houses and lacks implementation. Yasmeen Aftab Ali in her book, *a comparative analysis of media and media laws in Pakistan*, has come up with a set of suggestions which PEMRA may like to look into. They are:

- Approval of editorial policies of media houses by PEMRA
- Media policies must take into consideration the PEMRA Ordinance 2002 and Amendment 2007
- If an anchor violates the channel's editorial policy, he/she must be banned from joining another media house for at least 3 years just like cricketers are banned from playing cricket if they are found guilty
- Impositions of a hefty fine in case of going against its editorial policy or the banning of channel in case if serious violation
- There should be guidelines for appointing anchorpersons
- Anchors must be allowed to moderate only those programs in which they have a strong background otherwise there should be a co-moderator who has expertise in that area
- Experts of different fields must be invited to media channels to ensure emergence of a balanced and well informed public opinion

The Dilemma of Funding of Privately Owned Media

Funding of media is a matter of crucial importance in almost every society but Pakistani society is much more vulnerable to this dilemma for many reasons with the most vital of acquiring corporate gains. The pursuit of these gains comes into clear confrontation with the concept of safeguarding the national interest of the country.

In an interview with Mr. Rizwan Razi, he said that when there was the rapid launch of private media channels after 9/11, "the parameters and content guidelines were being sent by the funding entities. And I have seen recommendations being made in favor of some Ulemas to get their program started on channels. So, there were hidden sponsors" (Razi, personal communication, September 9, 2014).

Independent media, in most societies, gives a tough competition to government entities but with reference to Pakistani society, it is much more evident that media outlets tend to be critical of the government. One of the reasons for this is that media is funded from the unknown private financers which can be in the form of political parties or anti-government lobbies. As a result, what a common man watches on TV is not the projection of the reality but a fabricated version of it. "This is the reason that media in Pakistan is mostly critical to the government and there is a controversy between the government and the media" (Hussain, 2008).

Pakistani media in general and a powerful media group in particular have been in the limelight because of their sources of funding. A recent wave of discussion, regarding Geo vs. military clash, highlights voices coming from different segments of society claiming that Jung group promotes the agenda of the countries it receives funding from. "Pakistan Tehreek-e-Insaf (PTI) Chairman Imran Khan alleged that the Independent Media Group, which owns Geo TV, is a defaulter of billions of rupees and its owner, Mir Shakilur Rehman, is accruing funds from foreign countries to promote their agenda in Pakistan (Muhammad, 2014)."He further added that, "Mir Shakil also received funds from India for the '*Aman Ki Asha*' campaign to promote Indian interests in Pakistan" (Muhammad, 2014). However, such kinds of claims have been negated by the owner of that media group.

The hidden funding sources are subject to have their motives met in society and for that matters many mediums are employed such as NGOs and MNCs. Events are arranged to show the positivity they are striving to bring in the society but what a common man sees in the projection of reality is totally different from reality.

Threats to sovereignty and security are not just what a country faces on its borders but threat perceptions have been changed and unknown sources of funding of Pakistani media are a real threat to the sovereignty of the country. In these chaotic circumstances, what Pakistan really needs is a free and fair media to help identify the enemies with bad intentions rather than being an enemy of the state itself.

State Media: A Pivotal Puppet

State controlled media acts as a puppet in the hands of the ruling elite. It is almost everywhere in the world is a government's mouth piece. (Razi, personal communication, September 9, 2014) Pakistan is no exception to this matter. The only difference is that for a long time since the independence of Pakistan, state media has enjoyed the status of oneness but the previous decade has added its competitors to the scenario.

Today the focus of people has been shifted from state controlled media to other private channels. This is because they find state media praising the present government and the so called prosperity it has brought to the country all the time. It is satirical that on the one hand, the country is shattered with the wave of terror and on the other everything seems peaceful on PTV. That is ironic! Today, PTV standards are said to be declining. "The Senate special committee has criticized the falling standards of PTV news and general programs and their declining viewership as compared with private TV channels" (Burhan uddinn, 2011).

In a society like Pakistan where extremism is getting increased with every passing day, it is imperative to ask that has state media played its role of promoting nation building? The answer is certainly not very hopeful. PTV planners should develop campaigns instead of wasting time on defending their prime minister and ministers (Rauf, personal communication, August 30, 2014). State media should serve as a bridge between state and society and links of political communication should be established and they shouldn't only praise the ruling government but try to show the objective view which PTV has failed to portray until now.

A Comparative Study Based on Interviews

To highlight the areas where Pakistani media needs to work in order to come up with journalistic standards and to secure the national interest, it is imperative to incorporate the journalists say in it as they are an integral part of this system. Two interviews have been conducted in this regard. First features Mr. Abdur Rauf who is associated with Geo TV and the second interview is being conducted with Mr. Rizwan Razi who was previously associated with CNBC Pakistan and now with Boom FM. The interview responses have been shown in a comparative format below.

Questions		Responses	
		Mr. Abdur Rauf	**Mr. Rizwan Razi**
Q1	In the aftermath of 9/11, US media identified its enemy through its reporting of war on terror, have Pakistani media played its role in the identification of the enemy?	We probably have not been able to identify our enemy due to the certain interests of certain groups. Because our media is full of people who have their backing or basis somewhere else and who are sponsored by certain groups. It is because of that sponsorship we have been totally preaching what certain groups want us to preach. The power brokers in this country want people to be confused. Media also is controlled by certain groups. They have kept people confused.	In my opinion the making, shape, construction and evolution of Pakistani media is not intended or not been able to watch, craft and define national interest nor has it identified the enemy. Serving the national interest is the next level; this media can't even define the national interest. Pakistani media is not there to watch the national interest of the country.
Q2	How will you regard the content of media channels (talk shows in particular and media reporting in general) either sensational or responsible?	It is sensational in some respect because at the moment there is a war going on within the media groups because of the influence and sponsorship of certain very powerful quarters. I believe there are anchors that sensationalize things. Also, the players who have entered this field are not professional media men. They are businessmen and their interest is in making money and whoever pleases them they start playing their tune. I would say ARY is doing that. Express is doing that. There are certain small channels doing that. So responsibility, I think, not now.	The content is highly irresponsible which creates sensationalism. Firstly, tragedy is that PEMRA has the culture of nepotism and secondly there is a monopoly of rating system. Ratings are manipulated and maneuvering is very much there.

Q3	Do you think that the use of different terminologies by media such as shiddat pasand, inteha pasand or shaheed creates confusion among people?	I think these terms shiddat pasand, inteha pasand, shaheed basically are given to the media by powerful circles and media follows that quite strictly.	I have been a great critic of it. In same kinds of incidents, some people are being called 'Halak' and others 'Shaheed or Janbahaq.' It is such a dangerous thing upon which regulatory body should work. It is even much more dangerous when issues like terrorism, sectarianism and ethnicity are involved. Every person looks it from the local angle and develops a rift for other sect or Ethnic group. So mental disturbance and destruction is created in society.
Q4	What are the weak areas Pakistani media needs to work upon?	I think the basic flow or weakness here is that media has not been able to develop campaigns which influence people with respect to education, science & technology, women's rights, population explosions, good schooling, bad schooling and campaigning for good healthy culture and not been able to convince people that reading good literature is so important for a good nation.	We need to educate our people regarding economy to the fullest. Media has frivolous, baseless and fruitless discussions which create mental distortion and political discussions in which participants abuse one another and increase ratings. Media is not educating people. It should highlight the aspect of education.
Q5	Do you think that state media is playing its role as it should be? Is it fulfilling the concerns of nation building and national integrity or just playing as a puppet?	PTV has been most powerful media in our country for many years and still PTV has many channels than any other channel/ network. What role is PTV playing? Everybody is forced to pay a fee for PTV. No other channel is asking you to pay a fee. So, We should first of all catch PTV planners and the minister whosoever is in charge by the throat and ask them to at least spend time on teaching the people, airing campaigns instead of wasting time on defending their prime minister and ministers.	State media almost everywhere in the world is a government's mouth piece. So expecting something good from state media I think is wishful thinking. State media is just like that pigeon that stays in the cage even when the lock is opened.

Q6	How the content of media in form of talk shows or regular reporting can be made less sensational and much responsible?	We need to have educated people in the media. Look at the western world and the quality of people they are producing. We only want to bring forward faces which are pretty. We need education in the masses and the media. We also need to improve our rating system. And only then the situation will improve otherwise no.	We need to multiply the no. of content analysts and we need to produce PhDs in that field being appointed on good positions in PEMRA who watch every program and highlight its negative and positive points and notify the producer and anchor of that program that if the same negatives are repeated again, program will be off air and it should happen in actual. By creating such an example things will be streamlined.
Q7	How do you see the role of PEMRA in Pakistan?	PEMRA is totally useless because the government is so weak. Nobody knows in this country that who controls PEMRA. It seems as if PEMRA is not controlled by the government but by some other power. I mean our forces control PEMRA, our army controls PEMRA at the moment.	The main problem is with regulatory authority. If regulatory authority would have been doing its work, it would have come up with narratives in favor of national interest. Everywhere in the world regulatory bodies give narrative for national interest. PEMRA should give narrative of national interest to channels by defining a line what channels can and cannot discuss.

There are many areas of agreement and disagreement in both the interviews being conducted but the one thing upon which both the interviewees agree is that our media has not been able to identify the enemy of the state and is contributing to create a confused nation. Channels are not observing PEMRA's regulations and there is no check on them to do so. PEMRA needs to be revised to better implement its ordinance and regulations.

Conclusion

This study was set out to explore what role Pakistani media is playing regarding national security and what are the weak areas which may possibly affect it negatively. After examining media's role in national interest and national security, it is important to highlight that Pakistani media dwell in sensationalism, a culture of commercialism and a war to gain towering ratings than to safeguarding its national interest. Anchors sensationalize serious issues to make their program stand out and what's worse is that instead of discouraging this culture, almost all the media houses have adopted this tactic. There are many areas Pakistani media falls short of such as it;

- Has not been able to identify the enemy of the state
- Does not have responsibility and objectivity as its first priority
- Lacks projection of a national narrative
- Uses such terminology which creates confusion among masses
- Is short of giving awareness to people regarding national security
- Lacks campaigning for a good healthy culture

The media of major powers like US and UK is the world's most regulated media under FCC and OffComm respectively. But in Pakistan, the regulatory body PEMRA is not performing its duty accurately. Its regulations are just like clichés to Pakistani channels. Electronic channels violate them and there is no such measure which can make them accountable in its true sense. Pakistan Electronic Media Regulatory Authority lacks:

- A strong hold/ authority over news channels
- Implementation of its 2002 ordinance
- Defining and injecting narratives to media
- Content analysts who can analyze content being showed

National security is a sensitive issue and media can either promote it or demote it. The coverage of sensitive issues such as attacks on our strategic locations needs to be dealt with responsibility and objectivity. There are times which demand to protect national interest such as at the time of coverage of Bombay attacks and the media vs. military clash lately rather than reporting and digging facts.

On regional and global levels, wars are not fought just with weapons but with the non-traditional means of propaganda. In the post 9/11 world, our media has been failed in countering the propaganda staged by international media. Moreover, regionally, we have a constant threat not just on our borders but in form of propaganda warfare by our traditional enemy. What media needs to do is to respond to the propaganda in an equal manner keeping the national interest of the country supreme.

I cannot conclude without quoting **Walter Lipman**, "There can be no liberty for a community which lacks the means by which to detect lies." The point to ponder is that either it's sensationalism over accuracy or commercialism over credibility, irresponsibility over responsibility, subjectivity over objectivity, confusion over campaigning to educate people and a rush to breaking news over authenticity, who pays? The society and the people living in it! The question that I leave the readers with is: Who will take the responsibility and when?

Findings

- After 9/11, US media identified the enemy of the state in form of Islam vs. west and safeguarded and enhanced their national interest by supporting the Bush doctrine of war on terror but after 14 years of independence of media in Pakistan, it has not identified the enemy of the state. People are confused and media is adding up to their confusion.
- It has all become about taking the credit for breaking the news at its earliest without measuring the harm such news breaking can bring to the national goals and purposes.
- The culture of commercialism dominates world as well as Pakistani media. The channel or the program which has good ratings, advertisers invest there and maneuvering and manipulations with the TV ratings is a serious problem which adds up to sensationalism and needs to be revised.
- The elements of thrill while reporting needs to be minimized on security related issues. Statements of militant and terrorist organization should get no place in media coverage.
- Media's reporting in the past regarding 'Ajmal Kasab' and in present times regarding the accusations on military agency by a particular media group are harmful for Pakistan's image. There should be difference between accusation and reporting. Pakistani media should refrain from accusing without evidences and should be objective in its reporting.

- One of the many weak areas Pakistani media is suffering from is the use of confusing terminology. Under same incidents, one person is called shaheed or jabahaq and other is called halak. Who to call shiddat pasand, inteha pasand or fascist, is another illustration of this confusion. This needs to be settled.
- Anchors behave like commentators on TV screens and show biases. Taking sides doesn't suit media. Clarity of job is lacking and information sometimes is broadcast without authentication.
- State media in Pakistan is just government's mouth piece and is not fulfilling the need for nation building and national integrity.
- PEMRA is just an issue licensing authority. It has a framework and a code of conduct but the dilemma is that it has never been implemented from the very start.
- Covert and clandestine sources of funding of Pakistani media are promoting the interests of those who want to paint a certain picture. Some media professionals are of the view that there are international sources of funding of media while others call it within the country funding from certain groups which want to have a control over media to preach what serves their interests.
- The art of agenda setting is essential to media policy making. Pakistan needs a well-planned agenda in favor of state and the interests of its people.
- The right of media's freedom of expression cannot be denied but absolute freedom is nowhere in the world. The most regulated media in the world is of the US and the UK's. Media in Pakistan needs regulation in its true sense.

Recommendations

- Editorial policy makers of media houses need to develop a national interest and security narrative in compliance with defense and interior ministry and that should be reflective in the media coverage of events.
- As the international media uses 'framing' to peruse state's agenda, Pakistani media should employ framing in the identification of the enemy.

- To end the monopoly of a single rating company, government should introduce a parallel rating system to minimize the culture of sensationalism.
- It is suggested that government should control the sources of funding of media and should make a unit under PEMRA whose core purpose should be examining these covert funding sources and to discourage them.
- There should be an immediate implementation of PEMRA's content and other regulations and the channels violating them should be heavily fined and banned for a certain period of time especially on national interest and national security related issues.
- Media holds the power of persuasion. By employing that, Pakistani media should work on building a softer and positive image of Pakistan on both national and international levels. In a present internal and external security scenario, Pakistani news channels should review their editorial policies and should give input for nation building which is very much needed at this time.
- People have the right to know but when it comes to national interest, they need not to know everything. Media should define limits for reporting on issues such as terrorism, sectarianism and national security related issues.
- Media, rather than creating hype regarding national security, should encourage debates involving specialists and national security analysts to create awareness among common masses.
- Media should not consider itself the policy maker hence only the messenger of the policies made by institutions.
- When issues like terrorism, sectarianism and ethnicity are involved, media should focus on developing a collective set of terminologies to avoid hurting the sentiments of people.
- To minimize the culture of yellow journalism and sensationalism, it is recommended to have educated people in the media and content analysts on good positions in PEMRA. We need to come out of bringing 'pretty faces syndrome' on screens. Rather than that people with responsibility and good qualification should be encouraged.

- It is suggested that talk show producers should stop sensationalizing sensitive issues like terrorism, sectarianism, ethnicity, religion and national security related issues and promote logical debates to better handling such sensitive issues.
- Media should be concentrating on developing the society into a healthy society through campaigning. Our media should made efforts to make people understand that they need to fight poverty, illiteracy, sectarianism, rigidity and they need, first of all, to develop into a strong nation where they are educated and understand what is right and what is wrong.
- It is suggested that PEMRA as a regulatory body needs revision and it should be made an independent authority free from influence of government or any other entity. Guidelines can be taken from international media regulatory authorities such as federal communication commission and OffComm.
- Pakistani media needs to get rid of its biases and its unregulated practices. It should develop responsible practices, caution and regulation.

References:

1. Morgenthau, H., Thompson, K., & Clinton, D. (1948). *Politics Among nations*. New York: McGraw-Hill Education
2. Kellner, D. 9/11, Spectacles of Terror, and Media Manipulation: A Critique of Jihadist and Bush Media Politics. Retrieved July 20th, 2014 from *pages.gseis.ucla.edu/faculty/kellner/essays/911terrorspectaclemedia.pdf*
3. Abrahamian, E. (2003). The US media, Huntington and September 11. *Third World Quarterly,* Vol 24, No 3, pp 529–544, 2003. Retrieved June 08, 2014 fromwww.hks.harvard.edu/fs/pnorris/Acrobat/Abrahamian_Clash.pdf
4. Resse, S. D., & Lewis, S. C. (2009). Framing the War on Terror. *Journalism* , 10(6), 777–797. DOI: 10.1177/1464884909344480
5. O'Hanlon, E. (2003). Televised Media: Supporting the War on Terror and the Impending War on Iraq. *Ethics of Development in a Global Environment.* Retrieved July 18th, 2014 from *web.stanford.edu/class/e297a/Televised%20Media.pdf*
6. Manzar, N. (May 5, 2014). Media & national interest. Dawn. Retrieved July 17, 2014 from http://www.dawn.com/news/1104294
7. War on Terror – Part 2: Political Parties and Media in Pakistan. (2008, September 28). Retrieved from http://emmjay.wordpress.com/2008/09/28/war-on-terror-part-2-political-parties-and-media-in-pakistan/
8. Cohen, B. (1963). The press and foreign policy. NJ: Princeton University Press.
9. Hoodbhoy, P. (September 27, 2014). How TV dupes our public. *Dawn.* Retrieved September 28th, 2014 from http://www.dawn.com/news/1134573
10. Ministry of Interior, Government of Pakistan. (2014). *National Internal Security Policy – 2014 -18.* Islamabad. Retrieved from *nacta.gov.pk/*
11. Hussain,N. (2008) 'Role of Media in Internal Security of Pakistan', *Journal of Political Science,* University of the Punjab, Issue XIII, Spring 2008.
12. Orwell, G. (1946). Politics and the English Language. Retrieved May27th, 2014 from www.npr.org/blogs/.../Politics_and_the_English_Language-1.pdf
13. Buckingham, D. (2003). *QUESTIONING THE MEDIA: A GUIDE FOR STUDENTS.* Retrieved from www.amarc.org/documents/articles/buckingham_guide.pdf
14. Schoenfeld, G. (2011). *Necessary Secrets: National Security, the Media, and the Rule of Law.* New York: W. W. Norton & Company
15. Hussain, N. (2008). *The Role of Media in National Security: A Case Study of 1998 Nuclear Explosions by Pakistan.* London, UK: South Asian Strategic Stability Institute. Retrieved May 23rd, 2014, from www.sassu.org.uk/pdfs/Report-20.pdf
16. Jan, O., M. (May 13, 2014). Secret agencies, media and the national interest. The Express Tribune. Retrieved May 23rd, 2014 from http://tribune.com.pk/story/707502/secret-agencies-media-and-the-national-interest/
17. Ali, Y. A. (2012). *A Comparative Analysis of Media & Media Laws in Pakistan.* Lahore, Punjab: Sang-e-Meel Publications .
18. Eijaz, A. (2012). Media Affecting Upon or Affected By ForeignPolicy: The Case of Pakistan. *Journal of Alternative Perspectives in the Social Sciences, Volume 5 No 1,* 74-95.
19. Entman, R. (1993). Framing: Toward clarification of a fractured paradigm. Journal of Communication, 43(4):51{58, 1993.

20. PEMRA. (2012). *Pakistan Electronic Media Regulatory Authority (content) regulations*. Retrieved from http://www.pemra.gov.pk/pemra/
21. PEMRA. (2012). Code of conduct for media broadcasters or cable tv operators. Retrieved from http://www.pemra.gov.pk/pemra/
22. Muhammad, P. (May 18, 2014) 'Foreign funding': Geo/Jang Group defaulter of billions of rupees, claims Imran. *The Express tribune*. Retrieved July 15th, 2014 from http://tribune.com.pk/story/709896/foreign-funding-geojang-group-defaulter-of-billions-of-rupees-claims-imran/
23. Uddin, B. (December 30, 2011).*PTV and Private News Channels*. Retrieved July11, 2014 from https://groups.google.com/forum/#!topic/presspakistan/P1dYwcfPXd0

CHAPTER NO. - 7

THE HOMELAND SECURITY INITIATIVES FOR PAKISTAN.... A GRAND STRATEGY

The homeland security system is a concept born to prevent the threat of terrorism and all types of crime in the urban city structure of the developed world. The power and economy of the globalized world paved the way for safe cites to attain sustainable development through communication technologies and modern strategies. In post '9/11' scenario the world encountered with the threat of terrorism that forced the endangered countries to launch collective efforts against the growing threat of terrorism. The homeland security strategy is a multi-faced task to provide safety and security to our cities and sustain its development. It encompasses the basic principles of partnership, convenience, sharing network, transparency and follow-up. Pakistan is in dire need to opt such strategies of safety and security to sustain its economic development and its citizens' security. The study will discuss a proposed strategy of homeland security for Pakistan keeping its focus on both approaches security and safety. It also studies the U.S., U.K. and the Indian models of homeland security.

a) Homeland Security: Pakistan

Domestic Dimensions of Pakistan's Security

Security is the only conception through which, the national security problem can be approached. It consists of not only military, but also political, economic, social, humanitarian and human rights and ecological aspects. Security emerged as one of main concerns of the contemporary world but due to its complexity and seamlessness the concept remained neglected. The notion of security is conceptually weak or undefined that needs the support of other approaches as Realism and Liberalism. It is a difficult concept as compare to other concepts of social sciences as power, justice, peace, equality and freedom due to its ambiguity. It has to be defined under an area of concern rather than a precise condition. The element of security has been defined by W.B. Gallia as an 'essentially contested concept' due to its relational attitude with other factors and un-rationality of meaning and application. As Buzan explains that security is a relational phenomenon, it involves not only the capabilities, desires and fears of individual states, but also the capabilities, desires and fears of the other state with which they interact. He further explains that,

> "If the security of each is related to the security of all, then nothing can be fully understood without understanding everything. The reality of security interdependence is

unavoidable. Consequently, the only hope of defining manageable subjects for study that neither lose, nor succumb to, the vital sense of whole, is to find a hierarchy of levels within the holistic perspective. Each of these levels must identify durable, significant, and substantially self-contained features of the security problem". (Buzan & Rizvi, 1986: 5)

According to United Nations, security is a condition in which states consider that there is no danger of military attack, political pressure or economic coercion that they are able to pursue freely their own progress. Another perspective by Michael H. H. Louw of security is to include a nation's traditional defense policy and also the non-military actions of a state to ensure its total capacity to survive as a political entity in order to exert influence and to carry out its internal and international objectives (Malik, 2003:2-3).

Pakistan's domestic security concerns are defined by its vulnerabilities and threats to its existence due to its geopolitics, location, political instability, economics and history. It is multi-nation state comprising on four distinct nationalities; it is extremely difficult for them to agree on what is national? Therefore, Rizvi identifies three interlinked components of the state, ideology, the sovereignty/territory and the political institution to understand Pakistan's national security structure. He believes that many of Pakistan's security threats are from within the state. "Traditionally, the threat to Pakistan's security has been external. However, today it can be arguably asserted that the major threat to national security in Pakistan emanates more from internal sources rather than external"(Javaid, n.d.: 1). General Ashfaq Pervez Kayani, Pakistan's Chief of Army Staff has admitted in his policy speech in 2012 that internal threat is bigger than that of external threat (Ahmed, 2014: 5). There are three factors particularly contributing to her traditional domestic insecurities: 1) a narrow and weakly defined purpose of Pakistan in terms of the concept of Islamic states; the ideological base of Pakistan is threatened by the ethnic, cultural and lingual dominance of the four provinces. The successful Islamic ideology behind Pakistan movement cannot be sustained among the heterogeneous nature of people of Pakistan. 2) The absence of consensus on the evolution of national institutions, the failure of leadership and the political institutions in developing a common strategy to address all the security concerns. 3) The heterogeneous nature of the state that dominated its own area with weak federal connection.

Pakistan's Internal Security Issues after '9/11'

Pakistan inherited unsupportive domestic security dimensions and its geographical and the administrative division of the sub-continent, which further contributed into its traditional security challenges. In the current situation, Pakistan is facing large number of issues, which are constant threats to the security, including environment, narcotic trafficking, terrorism, separatists movements, disarmament, proliferation of weapons of mass destruction, human rights and gender issues (Pakistan's Internal Problems (all aspects), 2013:3). In post-'9/11' scenario, Pakistan confronted with the internal, non-traditional security threat by the Jihadist terrorists besides the traditional internal security threats of ethnic terrorism, nationalist terrorism, and sectarianism. Terrorism can be described as a non-traditional form of violence or a political message through violent acts that symbolizes a struggle against the establishment.

Pakistan paid heavy price due to its involvement in post - '9/11' US-Afghan war, initially Pakistan supported Taliban, later it was pushed to opt for anti-Taliban strategy to coup with the international community. The unanticipated Pakistan's policy aggravated the extremist Taliban-Jihadis and turned their violent tone towards Pakistan. "Even though Musharraf demonstrated wisdom by supporting U.S. war on Terror against Taliban in neighboring Afghanistan, the country continues to be caught in a very complex web of terrorist activity" (Javaid, n.d.:6). "9/11 was only one terrorist incident on the U.S. territory but inside Pakistan number of such incidents have been observed, which deteriorated the law and order of country. No part of the country is safe due to bombing and suicide attacks" (Afzal & Iqbal & Inayay, 2012: 196). The growing internal terrorism threats affected and changed the foreign policy of Pakistan and brought drastic disturbance in security structure of the state. Although terrorism was not a new phenomena for Pakistan but the intensity and diversity of the violent acts have been increased after the second Afghan war. Despite of Pakistan's efforts with the international community to curb terrorism, about six million Pakistani brains have been moved to safe and secure places as Canada, Australia, America, England, United Arab Emirate and South Africa due to the insecure environment and the violence of terrorism, it also damaged the social and economic activity of the state. "During the last ten years after U.S. invaded Afghanistan, Pakistan economy has suffered around 68 billion US dollars loss" (Hussain & Hyder, n.d.: 3).

"Talibanization promoted the culture of terrorism in Pakistan that includes ambushes on armed forces by Tehrik-e-Taliban Pakistan (TTP), target killing of religious, political and civil society figures, scholars and doctors, bowing up schools, offices of NGO's, bridges and pillions, suicide attacks in mosques and other important places. So far more than 35000 Pakistani have lost their lives in different incidents of terrorism including the loss of 5000-armed men" (Afzal & Iqbal & Inayay, 2012: 197).

Pakistan is passing through severe threats of terrorism that has paved the way for many violent acts of sectarianism and national terrorism in Baluchistan. It increased Pakistan's vulnerability and pushed a nuclear country to fall down. Pakistan supported all the UN resolutions in order to curb terrorism; it also signed ten UN anti-terrorist convections. The government of Pakistan also initiated many steps in order to prevent terrorism and violence, as follows:

- "Cooperation with the international coalition to curb terrorism
- Immigration control measures to keep a check on the movement of potential extremists
- Action against fanatical and extremist organization and groups
- Detention of Jihadi leaders
- Banning Jihadi organization
- Reformation and regulation of madrassas encompassing:
    - NOC(No Objection Certificate) for the opening of madrassas
    - The registration of madrassas
    - A review of their syllabi
    - A curb on the number of foreign students attending them
- Regulation of Mosques:
    - NOC required for opening a mosque
    - Political activities in mosque made unlawful
    - The use of loudspeakers restricted to the *Azan* (the call to prayers)". (Khan, 2013:44)

Nonetheless, Pakistan is caught in a wave of intense wave of terrorism after the "9/11" attacks, which are leading and providing more opportunities to different types of crime and radicalization especially in urban areas. Pakistan is in need for a homeland security strategy to provide an outlined plan against the increasing terrorist acts and crimes. This initiative can provide Pakistan with safe and secure atmosphere that can lead to prosperous economy and sustainability in the urban development.

b) Concept of Homeland Security

Origin and its Historical Evolution

Safety and security have always been one of the most important essentials of every state. In the ancient times, cites were designed and fortified in such a way to encounter any external and internal threat; similarly it is of great importance in today's world. In the modern stream line of development the world is moving towards rapid urbanization. The UN predict that 69% of the world population will be urbanized by 2050.The continuous technological advancements and globalization also contributed in the rapid process of urbanization and connected cites with each other. Therefore, a dire need for safety and security emerged in the new arena beside the economic and developed city infrastructure to ensure prevention policy against all types of crime and threats. The homeland security structure prevents all types of threat as the separatist's acts, ethnic and terrorist violence; cross border infiltrations, threats to human security, individual local crime, drug trafficking, arms smuggling, and the circulation of fake currency.

"Pre – '9/11', law enforcement criminal intelligence focus was on specific types of crime, such as organized crime, white-collar crime, or gangs"(Davis, Pollard, Ward, Wilson, Varda, Hansell, Steinberg, 2010: XIX). In post – '9/11' scenario, security and counter terrorism measures emerged as one of the most required strategy of every state. The continuous threat forced the world to shuffle their security policies. The U.S. policy makers concluded the approach of 'homeland security to address the large-scale terrorist acts. The unconventional terrorist's methods targeted the urban cites to destroy the economic activity of the state and intensified the damage with maximum civilian carnage.

"The ordinary criminal certainly short-term to achieve his goals,

> be it brandishing a knife in front of a mugging victims or using a gun in a bank robbery. But the purpose of the criminal act does not go beyond the act itself or the acquisition of money and other valuables. The terrorist act is different in that the violence employed is not only in pursuit of some long-range political goal but it designed to have far-reaching psychological repercussions on a particular target"(Riley, 2010: 12).

In the aftermath of '9/11' attacks, we can observe an obvious change in the security structure of United States that addressed both safety and security. Later, they also included in the homeland security, the federal, state, and the local government responses to disaster such as Huuicane Katrins. They expanded the concept of homeland security to include significant disasters, health emergencies, and others security concerns which threatens America, as its economy, the rule of law and government operations (Reese, 2013: 5). The Law Enforcement Agency (LEA), which is the key acting actor under the federal entities, evolved its information-sharing networks; the counter terrorism measures and it also included all the crime prevention approaches. "As these agencies seek to build intergovernmental partnership and sort out their respective roles and responsibilities for homeland security, they will have to work through historical conflicts concerning authority, jurisdiction, the distribution of federal aid, and the sharing of intelligence and other resources"(Holden, 2003: 2).

The structure of homeland security encompassed both the criminal intelligence and the information related to terrorism threats. The U.S. Homeland Security Department came into existence in 2002 as an immediate post -'9/11' affect that restricted Al-Qaeda's vicious intentions and their capacity to orchestrate another major attack against the homeland due to the counter terrorism efforts (Tankel, 2011: 3). While it was initiated in UK after the 2005 London blasts and the Indian followed the same response after the Mumbai mayhem on '26/11'.

Homeland Security Structure

The homeland security structure is prepared to protect, detect, and to respond to any type of aggression on its focused urban cities. Its brief structure includes counter terrorism measures, safeguarding infrastructure, surveillance, intelligence, and policing. Thus homeland security can be described as "a comprehensive and concerted exercise aimed to maintain the territorial integrity of a state and ensuring sustained peace, public order and rule of law in it" (FICCI, 2011: 3). According to the 2010 Bottom-UP Review (DHS), homeland security can be described as, which prevents terrorism, responds to any threat and recovers from natural disaster, customs enforcement and collection of customs revenue, administration of legal immigration services, safety and stewardship of the Nation's waterways and marine transportation system, as well as other legacy missions of the various components Department of Homeland Security (DHS). By summing all the definitions, Reese (2013) concluded the purpose of homeland security in the following points.

- The homeland security enterprise encompasses a federal, state, local, and tribal government and privet sector approach that requires coordination;
- Homeland security can involve securing against and responding to both hazards-specific and all hazards threats; and
- Homeland Security activities do not imply total protection or complete threat reduction.

The shift from safe city model to a broader homeland security strategy was due to the newly introduced element of terrorism in post Cold War scenario. Terrorism (in post-'9/11' scenario) can be described as the strategy that covers all the internal/external intentional violence prepared by sub-national non-state actors and transnational groups that has some specific political, economic, religious or social goals. "Terrorists aim to maximize loss to urban life and assets. With this in mind, they have shifted their focus from merely targeting government bodies to crowded areas in which they can wreak the maximum loss to life and property"(FICCI, 2011: 2). The terror attacks risk in a city depends on the population index of the city and the economic index because any attack can result in a substantial financial loss. Lastly the presence of critical infrastructure can also increase the terror hazards.

The traditional LEAs were concentrated on the crimes and its circumstances and how to prevent it, within the city, but in post – '9/11' the counter terrorism polices were added in the structural frame work of LEAs and a special funding was also introduced for the implementation of measures. "Lead our nation's effort to combat terrorism at home an abroad by analyzing the threat, sharing that information with our partners, and integrating all instruments of national power to ensure unity of efforts"(Olsen, 2012: 6). The homeland security strategy is based on the state and local law enforcement agencies, combined efforts for the prevention of terror attacks. They are engaged in initiatives to equip, train and provide "interoperable" communication support first responders; identify and develop plans to protect critical infrastructure; assessments of threats and vulnerabilities, investigation of terrorists and terrorist organization; and implement protocols and procedures for controlling a mitigating damage from a terrorist attack (Holden, 2003: 2).

> "Counter terrorism efforts intersect with traditional policing efforts along many dimensions. These include personnel allocation, recruitment, retention, training, internal communication, external communication/information sharing, technology, community outreach, resources and other crime areas. For example community policing is not only affected by counter terrorism efforts (resource allotment and prioritization of efforts) but also serves to enhance counter terrorism efforts (using local knowledge to under-stand the actual threat in a community among its members"(Davis, Pollard, Ward, Wilson, Varda, Hansell, Steinberg, 2010: 7).

It has been evaluated that the terrorist acts and local crimes run in the same circle against the law and order. Both are non-legitimate acts of violence. Terrorism has deeper influence and physical impairment while crimes have limited social disfigurement. They also require the same resources in order to conduct their transgressions, as a false identification, travelling documents, operators, transportation network, and counter surveillance techniques. The safety and the security intentions have been gathered under the broad itinerary of homeland security. They improved the sharing of intelligence information between the local, state and federal law enforcement agencies. As response to the '9/11' incident and under the homeland security of U.S., the LEAs were created and activated according to the security requirement of the state. Special homeland security initiatives have been introduced to implement the counter terrorism measures. They recognized the need to expand the data collection, analysis, and synthesis of information. They also encouraged the sharing network and created fusion centers between the local and federal LEAs to encounter the local crimes and the threats of terrorism.

> "A collaborative effort of two or more agencies that provide resources, expertise, and information to the center with the goal of maximizing their ability to detect, prevent, investigate and respond to criminal and terrorist activity. Fusion also turns information and intelligence into actionable knowledge. The primary components of a fusion centers are situational awareness and warnings that are supported by law enforcement intelligence, derived from the application of the intelligence process, where requirement for actionable information are generated and information is collected, integrated, evaluated, analyzed, and disseminated"(U.S. Department of Justice Programs, Global Justice Information Sharing Initiatives, 2006).

Another important factor is that these sharing networks and fusion centers have been successfully working with the adaptation of high technology and special funding by the federal entity. It also encouraged the LEAs to expand its structure behind the states' territory and build it ties with the multiple regional stakeholders within the large urban areas. They further advised for regional cooperation to increase preparedness, coordination, and integration, create fusion centers and to improve the sharing network at the regional level. The U.S. counter terrorism professional are working with their European counter partners to track the travelling to Pakistan and other suspected countries, and in the wake of Mumbai Attacks 2008 U.S. has also increased its counter terrorism collaboration with India to dismantle the terrorist networks in Pakistan. U.S. is also in process of building security ties with Bangladesh and Nepal and the Gulf states in order to prevent terrorism in the region (Tankel, 2011: 5). However, the process of regionalization and sharing will broaden the view of homeland security structure that can be explained as follows:

- Encouraged the adaptation of a regional approach to counter terrorism homeland security preparedness and response.
- Greatly expanded the size of the network for counter terrorism and criminal intelligence and information sharing.
- Helped formalize LEAs counter terrorism intelligence and analysis functions.
- Broadened the mission from terrorism focused only to all crime/all hazards approach (Davis, Pollard, Ward, Wilson, Varda, Hansell, Steinberg, 2010: 56).

Safe City Model

In the globalized world of today, everyone is seeking for a peaceful life in a city without crime and threats. Crime and violence are not new phenomenon but peace is fading away from our lives. "Cites are violent due to their profound divisions, both spatially and socially, which presents a particular model corresponding to anew temporality of globalization and the society of information" (Boisteau, 2006: 21). The concept of safe city initiated by the end of 1980s, the innovation came from Toronto in Canada, where this important experiment of urban safety has been carried out, which introduced new ways of tackling safety problems, city functions and the way citizens uses its spaces and facilities (Planning Urban Design and Management, 2006-2007: 10).

The rapid urbanization of our cities requires a new structure of safety to reduce the crime index. "Massoomeh *et.al.* (2011) contended that worldwide concern about population growth and rapid urbanization will continue to address crime as a social problem. Becker in Erdal (2004) pointed that crime become more important during the last forty years that is discussed all over the world and safety becomes a central dimension for contemporary debate on urban sustainable development (Ceccato & Lukyte, 2011)" (Shamsuddin & Hussin, 2013: 225). Therefore, the implantation of the safe city concept becomes prior need of every developing country. It provides a critical component of the overall smart city paradigm in the new millennium of the global urban growth. It has been initiated to maximize safety and to reduce crime in the society. A city free from all the physical threats, social disorders and polluted environments a community-based, public partnership led by the local law enforcement agencies. The safe city model has several broad objectives in order to attain the maximum safety. Specific goals of safe city are to,

- Strengthen the ability of law enforcement to lead local safe city initiatives;
- Reduce crime by implementing situational crime prevention techniques, communication technology, and other targeted solutions;
- Identify ways that safe city partners can effectively identify, prioritize, analyze, and prevent crimes;
- Increase collaboration among internal partners; and

- Create a formal infrastructure for partnership with the law enforcement, neighboring businesses, and community partners (LA Vigne & Owens & Hetrick, 2009: 32).

While the European urban planning and design management follows three basic approaches of urban safety. These approaches grantee environmental safety and urban development, the first approach deals with security, mainly in terms of law enforcement through the use of rules and police. The laws regulate people's action and the police forces watch whether laws are followed and respected by the people. Secondly the crime prevention approach in social terms, it acts to reduce the conditions of disadvantage and poverty, which are considered as the major factors in generating criminal and anti-social behavior in the society. The third approach promotes environmental crime prevention. It deals with all those factors in the environment, which can somehow affect a criminal act to take place. (Planning Urban Design and Management, 2006-2007: 9).

The implantation process of safe city is developed through four major components, (1) identifying key partners and garnering support from local Law Enforcement Agencies. Partners are the most important supporting frames to develop and sustain a comprehensive and viable crime prevention strategy. The Law Enforcement Agencies are considered the most important supporting partner, because it is the most logical entity to lead the safe city program. It has unique expertise in crime prevention and also has access to the crime databases and the crime reduction resources. The safe city implantations require a strong leadership "you need someone with the publically endorsed agenda and political power to push and support safe city investment,…if you don't have strong leadership you cannot bring a project such as this forward"(Life Talk, 2012:11). (2) Engaging key partners and formalizing safe city, setting expectations and limitations of the model identifying its challenges and resources and how the community will respond to the structured partnership. "One of the largest challenges is not the technology itself, but how you integrate all of it. These applications are very complicated, encompassing multiple technologies, multiple vendors and integrations". (Life Talk, 2012:11) (3) We have to identify and implement targeted solutions in the community, involving the community in the safe city program and determining their readiness to ensure security. We have also to identify the issues and challenges that can affect the program and preparing strategies. (4) We have to manage the safe city program, assets protection, and staff to work with the Law Enforcement Agencies to create an inventory of potential partners.

The shift from the concept of safe city towards a safe and secure city (homeland security) is explained through the following,

- "First, as LEAs counter terrorism efforts continue to move towards a fusion center model, large urban areas are starting to use information technology to organize virtually and to share information.
- Second, the renewed focus on counter terrorism and homeland security have served as a catalyst to promote technology adaption (e.g., IT system, software, camera systems) by fusion centers. An advantage is that this allows for better access to information, better linkage of network participants, and leveraging of technology to facilitate identifying the nexus between different types of criminal activity and potential terrorist-related activity.
- Third, LEAs have tapped into existing infrastructure and networks to help build their local intelligence functions.
- Fourth, fusion centers have helped centralized and formalize information exchange among LEAs and other network participants within a region.
- Fifth, homeland security and the Urban area Security Initiatives funding focuses on enhancing regional preparedness, with a key goal being to encourage a regional, multi-jurisdictional approach and improve coordinating among multiple stakeholders and different sizes of stakeholders within a region" (Davis, Pollard, Ward, Wilson, Varda, Hansell, Steinberg, 2010: XX).

c) Pakistan and the Homeland Security Initiatives

Pakistan's security structure is different from the discussed states and their homeland security initiatives. United States, United Kingdom and India opted for homeland security initiatives after some noticeable terrorist attacks, but in case of Pakistan, it possesses embedded extremism in its structure. The distinct element of religious extremism paved the way for anti-western sentiments and terrorism. Secondly Pakistan's own administrative decisions; political instability and their continued problematic relations with India further enhanced their vulnerability. 1979's Afghan-Soviet war and their undue involvement in War on Terror pushed the state into burrow of terrorism. India also possesses the elements of religious extremism and anti-western sentiments but its political solidity and economic development stabilized the state against the emerging threats. However, terrorism is an extending phenomenon; we cannot consider it limited to Pakistan. For more than a decade after '9/11', U.S. and other western states are on war with Al-Qaida and terrorism; they are still at threat even after the death of Osama Bin Laden. Al-Qaida's anti-American/anti-western ideology has its roots in many third world states. "These groups are from an array of countries, including Yemen, Somalia, Nigeria, Iraq and Iran" (Olsen, 2012: 1). The author further explains that these groups are in different states take directions from Al-Qaida-Pakistan. This argument is not a rational analysis of the situation; these are different transnational groups, allies of each other, with their own masters, the only similar thing that is the anti-western interests. According to the National Security of America (2010) that Al-Qaida's core in Pakistan remains the most dangerous component of the larger network, but we also face a growing threat from the group's allies worldwide. While the United Kingdom National Security (2008) strategy underlines Afghanistan, Iraq and Pakistan as the terrorists' safe heavens. It further reveals that the new terrorist threat is not a single organization but a wider network of affiliated groups.

The U.S. homeland security strategy has been based on the Homeland Security Act of 2002. The Act outlines the structure of homeland security implementations; it defines the responsibilities and guidelines of Homeland Security Department. It addresses the information analysis, infrastructure protection, transport security, emergency response, coordination, airlines war risk, federal workforce, and the technological development in order to support the homeland security, etc. (Homeland Security Act of 2002). The homeland security department also endorses its bilateral cooperation with India. The U.S.-India 2010counter terrorism cooperation (Memorandum of understanding) allows U.S. to facilitate capacity building and training in India, they are working to enhance,

- The investigative skills;
- Cooperation between forensic sciences and laboratories;
- Mutual investigative assistance;
- Capabilities to act against money laundering, counterfeit currency, and

terrorist financing;

- Trading best practices on mass transit and rail security;
- Interaction between Coast Guard and Navies on maritime security;
- Expertise on port and border security;
- Training between specialist counter terrorism units, including the National Security Guard, and their U.S partners.

In 2011 the United States and India formally launched their Homeland Security Dialogues, to enhance the initiatives for homeland security and counter terrorism measures (U.S-India Homeland Security Cooperation, 2013: 14). While the homeland security approach in United Kingdom is different from United States, "the United States has embarked on a wholesale reorganization of its domestic security and border protection institutions, European countries have largely preferred to work within their existing institutional architect to combat terrorism and respond to others security challenges and disasters, both natural and man-made" (European Approaches to Homeland Security and Counter terrorism, 2006: 2). None of the European countries including United Kingdome have an equal department to the U.S. Department of Homeland Security. The responsibilities of the adopted strategy of homeland security and counterterrorism measures are divided among different ministries. They have integrated the counterterrorism measures and preparedness programs into the existing security structure. The European homeland security is a matter of attraction to United States, and the U.S. policy makers suggests for greater U.S.-European cooperation in the area of homeland security in order to guarantee security on both sides of Atlantic.

However, in Pakistan, it is hard to trace the implemented homeland security structure, although there have been many primary initiatives of homeland security. Pakistan has been focusing on both military and non-military options to combat terrorism. It has promulgated several amendments in the existing legal framework dealing with anti-terrorism measures. The Government of Pakistan announced the Anti-terrorism Act (ATA) in 1997, which was later, amended in 2002 (The Anti-Terrorism-Amendment Ordinance 2002). The Ordinance added the power of targeting the entire terrorist network with severe punishment up to death penalty. The Ordinance was further amended in November 2002 by giving the authority to police to hold a suspect for up to 12 months without filling any criminal charges. In 2004 another amendment was made in ATA 1997 which gave the right to appeal to the victims against the judgment of Anti Terrorist Court (ATC) in the High Court within thirty days. By 2005 they further modified the ATA 1997, in which they increased the penalty years to life imprisonment, and the also enhanced the ATC powers. In 2009, the government issued Anti-Terrorism (Amendment) Ordinance in which the establishment permitted the extra-judicial confession before ATC, and also extended the remand period from 30 to 90 days. The Pakistani government kept on enhancing the legal framework to counter terrorism; in June 2011 the government promulgated the Regulation Action in Aid of Civil Powers, the 2012 Investigation for Fair Trial Bill, and in 2013 the parliament passed the Anti-Terrorism (Amendment) Bill. Moreover the Parliament passed a bill in March 2013 for the formation of a new independent body, National Counter-Terrorism Authority (NACTA) (Ahmed, 2014: 10-11),

> "With functions 'to receive and collect data/ information/ intelligence, and disseminate and coordinate between all relevant stakeholders to formulate threat assessment; to formulate comprehensive counterterrorism and counter-extremism strategies; and to establish links with the international entities for facilitating cooperation in areas related to terrorism and extremism" (Ahmed, 2014: 11-12).

According the official website of National Counter-Terrorism Authority (NACTA), the Authority shall have the following functions, namely: -

(a) to receive and collate data or information or intelligence, and disseminate and coordinate between all relevant stakeholders to formulate threat assessments with periodical reviews to be presented to the Federal Government for making adequate and timely efforts to counter terrorism and extremism

(b) to coordinate and prepare comprehensive national counter terrorism and counter extremism strategies, and reviews them on periodical basis;

(c) to develop action plans against terrorism and extremism and report to the Federal Government about implementation of these plans, on periodical basis;

(d) to carry out research on topics relevant to terrorism and extremism and to prepare and circulate documents;

(e) to carry out liaison with international entities for facilitating cooperation in areas relating to terrorism and extremism;

(f) to review relevant laws and suggest amendments to the Federal Government; and

(g) to appoint committee of experts from Government and non-Government organizations for deliberations in areas related to the mandate and functions of the Authority.

The official website also prescribed the following powers that the National Coordinator shall have, namely: -

(a) to execute the policies and plans approved by the board and instructions issued by the Federal Government:

(b) to prescribe terms and conditions of the employees and grant additional allowances or any other incentives;

(c) to have full financial and administrative power for effective administration of the Authority, as approved by the Board;

(d) to engage any person or entity on contract basis to carry out assignments for the consultancy in accordance with acclaimed best practices;

(e) to establish administrative structures at the field level for efficient implementation and accessibility of the Authority;

(f) to submit quarterly progress reports to the Board on the Financial and functional aspects of the Authority;

(g) to perform such other functions as may be delegated by the Board;

(h) to undertake any other assignments given by the Board in the respective fields; and

(i) to produce periodical journals relating to counter terrorism and counter extremism issues (National Counter Terrorism Authority Pakistan, Source: http://nacta.gov.pk/).

On October 20, 2013, the President Mamnoon Hussain promulgated Pakistan Protection Ordinance (PPO). The new ordinance has provisions with high security prisons, separate police stations for specified crimes with federal prosecutors to argue cases in special federal courts. This also includes preventive detentions and enhanced punishments and such offences, as designated under the law, as it has a provision to conduct trials out of the jurisdiction of where the crimes were committed. The law assigned powers to police officers and the armed forces to arrest, without warrant, any person who has committed a scheduled offence or against whom a reasonable suspicion or credible information exists that he/she has committed, or is about to commit any such act or offence. (Protection of Pakistan Ordinance promulgated to fight terrorists, criminal elements: Zafar, 2014). However, PPO was highly criticized by the national and international community, I. A. Rehman stated that the legislation (PPO) is not needed. Many of its objectives can be secured through the Anti-Terrorism Act (Mujtaba, 2014: 2). Dr. Azhar Nadeem stated in his article on the counter terrorism front, the government came up with the Pakistan Protection Ordinance 2013, with a view to giving teeth to the law enforcement agencies (2013).

Analytically Pakistan is well aware of the internal threat consequences and has managed to draw the initial lines to combat terrorism and to how to deal with the dominating threat to homeland security. "Although the anti-terrorism legislation in Pakistan (since 1997) is a "bold departure from the normal legal system", (Kennedy: 390), it has proved insignificant to prevent terrorism" (Ahmed, 2014: 12).The state failed to undertake any speedy trials terrorist, less successful to ban the militant organizations, which eventually proves the weakness of the state and its rit. Dr. Azhar Nadeem in his article emphasizes on the national consensus to eradicate militancy and to come up with a pragmatic counter terrorism strategy (2012).

d) The Purposed Homeland Security Initiatives for Pakistan

Homeland security encompasses several dimensions of national focus and relates to safeguarding the internal environment of a country from disruptive activities that can potentially lead to social disorder (Homeland Security in India, 2010: 11). We have discussed Pakistan's internal security dimensions for security and its dire need for planned strategy that can define all the domestic security initiatives in terms of security and safety. The following hierarchical structure is the basic outline for Pakistan's security concerns, components can be defined through the following:

1) Border infiltration: Pakistan possesses multiple types of boundaries, frontiers and territories with different countries. Some of these boundaries are un-demarcated and un-delineated, and the others include ceasefire lines, lines of control, lines of actual contact and working boundaries. This demographic boundary structure gives rise to all type of cross-border security challenges. "Pakistan inherited boundary disputes with almost all of its neighbors, but succeeded in settling some of them. A number of claims and counter-claims remain with at least two neighbors, India and Afghanistan, as do a number of associated cross-border transit issues" (Neill, 2010: 6). The strategy has to focus on the element of border infiltration especially on the emerging issues from Pakistan-Afghan border dispute and the Indian influence in Baluchistan. In a meeting between the U.S. Central Command Chief, General John P. Abizaid and President Musharraf, they shared the evidence that infiltration was taking place from Afghanistan and about the involvement of foreigners in creating trouble in Baluchistan. Islamabad has its serious concerns over the location of six Indian consulates in cites along the Pakistan border, and the Indian assistance to build the Afghan army (Zeb, 2006: 71-72).

2) Counter terrorism, terrorism is premeditated use of threat of use of violence by individuals or substantial groups to obtain political, religious, or ideological objectives through intimidation of a large audience (Arce& Sandler, 2005: 183). Since '9/11' Pakistan is under serious threat of terrorism and NECTA is responsible for all the counterterrorism measures that can handle the situation. Besides the element of terrorism Pakistan is also facing a separatist movement in Baluchistan. "Islamabad is reported to have provided proof to the Afghan President of the role of the Indian consulates in Jalalabad, Kandahar, Mazar-e-Sharif and Heart in fueling violence in Pakistan especially Baluchistan" (Zeb, 2006:70). Both aspects are highly matter of concern for the security structure of Pakistan.

3) Intelligence, the main aim of intelligence in homeland security structure is gathering and sharing the collected information to pre-empt any terrorist threat and crime. Intelligence in internal security management has various components like intelligence collection within the state, trans-border intelligence collection and the use of critical technology for the collection of intelligence specifically required for the internal management (Homeland Security in India, 2010: 14). The other main important aspects are communication, sharing network and the creation of fusion centers where they can share the data to encounter any threat.

The safety components can be defined through the following:

1) Critical infrastructure, It includes both public and privet assets that are strategically important to a state's interest, such as airports, industrial installations, national monuments, energy supply pipelines, nuclear and conventional power plants etc. "There is an immediate requirement for conducting a National Risk Assessment to identify and map the overall risk and vulnerability of the assets to any threat. The next step will be the development and maintenance of adequate counter-terrorism and emergency plans, including programs for providing awareness training to the security staff of such facilities"(Homeland Security in India, 2010: 15).

2) Police and Paramilitary Modernization; A well-trained, well-equipped and dedicated police is the basic requirement of every state to carry out internal security measures. In case of Pakistan, the police face the crises of legitimacy. Due to the colonial legacy and the political misuse of police, the public perceives the police as cruel, callous and corrupt. Secondly, the police suffer from the resource inadequacy and no information sharing system (Nadeem, 2002: 214-16). It is facing a critical down casting, the public has no confidence on the police, it lacks the operational and professional autonomy, and it has been always detracted and misused. "It is important to note that the professional autonomy of the police has been an anathema to successive government, eager to use them for harassing political opponents, winning elections and achieving extralegal objectives" (Nadeem, 2013: 1). Therefore the police and paramilitary modernization process considered as the focus area to strengthen the homeland security apparatus.

> "Apart from building national consensus to fight the forces of extremism with full state might, political parties must commit to providing adequate resources for the institutionalization of community policing involving effective public participation through various techniques of consultation, a balanced media policy revival of alternative dispute resolution mechanism, and professional conferences of police officers at the national, provincial and district levels with meaningful agenda. This can go a long way in preserving the rule of law. Last, but not the least, adhoc measures like entrusting police functions to paramilitary forces, like the Rangers and the army, need to be given (Nadeem, 2013: 1).

3) Safe city surveillance, the presence of CCTV cameras makes people more disciplined and also provides deterrence against the local crime. Especially it is required in the key events, where more chances are available of terrorism and crime. "The video surveillance now enables the government to monitor and control the quality of utility service providers' work. Building control systems allow for tight control over personal activities, leading to more accurate accounting and technical supervision (Safe city Moscow, n.d.: 2).

e) National Security Policy and Homeland Security

The National Security Policy is considered as, the nation's plan for the coordinated use of all the instruments of state power non-military as well as military, to pursue objectives that defend and advance its national interest (Doyle, 2007: 624). Providing security for the nation and for the citizens is the prime responsibility of every developed state in the previous and the current arena. The world observed shift in the global security structure in post-Cold War scenario. The opposition between two power blocs has been replaced by more complex and unpredictable set of relationships and non-traditional threats. The complexity of the globalized world further enhanced by the new economic trends and the communication advancements. Thus the entire change pushed the world into a more complex web of security concerns as, the international terrorism, weapons of mass destruction, conflicts and failed states, pandemics, and transnational crime (The National Security Strategy of the United Kingdom, 2008: 3). The U.S. National Security of 2013 also admits the change in international order since the end of Cold War, "The United States must prepare for a multilateral world where, while retaining our military, economic, and cultural preeminence, we may be challenged by both allies and adversaries". "President Barack Obama Administration's 2010 National Security Strategy states that strategy is meant to recognize "the world as it is" and mold it into "the world we seek" (Reese, 2013: 6).

Internal security emerged as one of basic needs of every developed state. Previously the national security strategy was considered to protect the state and its vital interests from any external threat. Over the recent decades the conception of national security extended and included the internal security concerns and their underlying drivers. The observed way of thinking among the Americans' public opinion' in the recent century with respect to fighting wars is probably also true, in some degree of all nations as they move toward more stable and affluent internal economies and by establishing the law and order situation within their borders (Davis, 1964: 83). Therefore the U.S emphasizes on strength at home and shaping the international order accordingly to meet their challenges of the time. They mainly defined four enduring national interests, security, prosperity, values, and international order. In the pursuance of security initiatives the U.S. policy continued to adopt integrated homeland security efforts seamlessly with other aspects of our national security approach. According the 2010 National Security Policy of United States that at home, it is pursuing strategy which is capable of meeting the full range of threats and hazards to their communities.

> "To prevent acts of terrorism on American soil, we must enlist all of our intelligence, law enforcement, and homeland security capabilities. We will continue to integrate and leverage state and major urban safe area fusion centers that have the capability to share classified information; establish a nationwide framework for reporting suspicious activity; and implement an integrated approach to our counterterrorism information system to ensure that the analysts, agents and officers who protect us have access to all relevant intelligence throughout the government" (National Security Strategy, 2010: 20).

On the other hand the United Kingdom National Security Strategy of 2008 also emphasizes on the importance of security, as follows:

> "The complex interdependence of the threats, risks and drivers of insecurity, in an increasingly interconnected world, is in itself a powerful argument for a single overreaching strategy for national security. This strategy focuses on the full range of issues, pulling together expertise and experience across government to produce a coherent response to the global security challenges of the twenty-first century" (P: 24).

While the Indian National Security strategy prioritizes the external security threats after the '26/11' attacks on Mumbai, an obvious shift has been appeared in the Indian security strategy. Since 1947, India pursued a grand strategy focused on preserving political unity aimed its bewildering diversities and potential rifts, protecting the nation's territory from internal and external threats, and realizing the economic development that would transform the country into a genuinely great power (Joshi & Mitra, 2014: 2). According to the Indian current National Security situation, India is centering on the challenges as, defense of national territory over land, sea, and air, borders security, off shores from religious, language, ethnic, or socio-economic dissonance, to promote harmonious relationship with the other regional players (China and Pakistan) that supports Indian national interests; and to contribute to international stability. However, the Indian counter-terrorism strategy is based on the constitutional values and political consensus. According to their summarized security strategy they emphasizes on the periodic net assessment of terror groups, patterns trends and finances. They also believe that the strategy will be enhanced by effective policing and collaboration among intelligence agencies, inter and intra–agency coordination, justice system, international cooperation, and civil society participation (Towards a National Security Strategy for India, 2011: 7).

On the other side, the study focused area of Pakistan's internal national security strategy is elucidated under the National Internal Security Policy - 2014-18 (NISP). Pakistan has been dominated by non-traditional threats of extremism, terrorism and militancy. NISP is the first ever-internal security strategy of Pakistan; its defined vision is to create a safe environment where people of Pakistan can live in harmony and freedom under the constitution of Pakistan. The policy is based on the principles of mutual inclusiveness and integration of all national efforts, including three basic components: 1) dialogue with all stakeholders, 2) isolation of terrorists from their support systems, 3) enhancing deterrence and the capacity of the security apparatus to neutralize the threats to internal security of Pakistan. The policy defines the implementing structure for the security strategy through integrated efforts with institutionalized monitoring structure under a democratic leadership.

NISP also chalks down the required policy framework for the implementation of this strategy, the soft component is the primacy of the political process and supported by Comprehensive Response Plan (CRP), including dialogue/ national narratives, infrastructure development, rehabilitation, monitoring through NACTA, reconciliation, reintegration, and legal reforms. The hard component is the Composite Deterrence Plan (CDP), in which integration of specialized skills and capabilities are required to combat traditional and non-traditional threats. They also believe in restructuring and building the capacity of the national internal security apparatus and creating efficient and coordinated institutions. NACTA will be the key operating authority to deal with all the required measures to counter terrorism. It will also liaise with the international actors for fostering cooperation to enhance anti-terrorism and anti-extremism channels (National Internal Security Policy, 2014).

Conclusion

The initial structure of a city was designed with the safety measures to encounter all types of crimes but in post-'9/11' terrorism scenario, the safe city model included safety and the security measures as well. The initial steps of homeland security were taken in U.S., than U.K. and after the Mumbai mayhem; they were initiated in India with the U.S collaboration. The concept is defined under different perspective; every state defines it according to its security concerns. Therefore the homeland security is defined with respect to the internal threat perception and the national interest of the state.

Pakistan's dilemma initiates from the insecure birth of new Muslim state surrounded with two hostile neighbors. The external insecurity paved the way for Pakistan's involvement in Afghanistan-Soviet War and the War on Terror, which laid down serious implication on Pakistan's internal security structure. It introduced terrorism, Kalashank of culture; it damaged the economic sector and raised the hypes of the separates' movements. However, Pakistan's security initiatives paperwork is appreciable but the actual impasse is the political instability and the clash between the civil-military leadership that has no national consonances on the matter of internal security. Secondly, the other damaging factor is lack of implantation, despite of being advised, and no actual information sharing occurs, no coordination is found among the Law Enforcement Agencies. Due to the unrealistic and unpractical approach of the governing bodies none of the LEAs are able to respond.

Reference

1. A Strong Britain in an Age of Uncertainty: (2010). *The National Security Strategy* , 09-36.
2. Afzal, S., Iqbal, H., & Inayay, D. M. (2012). Trrorism and Extremisn as a Non-Traditional Security Threat Post 9/11: Implications for Pakistan's Security. *International Journal of Business and Social Science, 03* (24), 194-203.
3. Ahmed N. (2014). Pakistan's Counter-terrorism strategy and its Implication for domestic, regional and international security. Fondation Maison des sciences de l'home.
4. Arce. D. & Sandler, T. (2005). Counterterrorism: A Game-Theoretic Analysis. Saga Publication.
5. Aviation Security: 10 Years after the 9/11 Attacks. (n.d.). Alpa Issue Analysis.
6. Boisteau, C. (2006). *Building Communities: Urban Planning And Security Polices.* ENAC.
7. Buzan, B., & Rizvi, G. (1986). *South Asian Insecurity and the Great Powers.* Macmillan Press Ltd.
8. Current National Security Situation. (n.d.) India. Retrieved from: http://fas.org/irp/nic/battilega/india.pdf
9. Davis, B. V. (1964). Principal Political Consideration in the Formation of National Security Policy. *The Western Political, 17* (03), 82-83.
10. Davis, L. M., Pollard, M., Ward, K., Wilson, J. M., Varda, D. M., Hansell, L., et al. (2010). *Long-Term Effects of Law Enforcement's Post-9/11 Focus on Counterterrorism and Homeland Security.* NCJRS.
11. Doyle, R. (2007). The U.S. National Securit Strategy: Policy, Process, Problems. *Wiley on behalf of the American Society for Public Administration.*
12. European Approaches to Homeland Security and Counterterrorism. (2006). *Foreign affairs, Defense, and Trade Division.*
13. Gupta, A. (2011). A National Security Strategy Document for India. *Share on live Journal* .
14. Glaeser, E. L., & Shapiro, J. M. (2001). Cities and Warfare: The Impact of Terrorism on Urban Form. *Harvard Institute of Economic Research*, 01-30.
15. Holden, Gwen A. (2003). *Building a Homeland Security strategy: State and Local Law Enforcement on the Line.* Washington D.C.: University of Pennsylvania, Jerry Lee Centre of Criminology, Forum on Crime and Justice.
16. Hilali, A. Z. (2001). India's Strategic Thinking and its National Security Policy. *Asian Survey, 41* (05), 737-764.
17. Homeland Security Act of 2002. (Nov. 25, 2002). United States of America.
18. Homeland Security in India An Overview. (2010). *KPMG.*

19. Homeland Security 2011, Building safe and secure Indian cities - a perspective. (2011). *FICCI.*
20. Homeland Security - Creating Safe and Secure Cities in India. (2012) Retrieved from http://blog.ficci.com/crime-safety-homeland-security-india/554/
21. Homeland Security 2013, Safe and Secure Nation - Procurement transformation in police forces. (2013). *FICCI.*
22. Hyder, S., & Hussain, S. (n.d.). Terrorism and Migration: A Case Study of Pakistan.
23. Javaid, U. (n.d.). Terrorism: Major Threat to Pakistan's National Security. Lahore: pu.edu.pk/images/journal/pols/currentissue-pdf/umbreen.pdf.
24. Joshi & Mitra. (2014). National Security: The Need for a Doctrine. India: Observer Research Foundation.
25. Khan, D. H. (WInter 2013). Pakistan's Contribution to Global War on Terror After 9/11. *IPRI Journal, 13* (01), 37-56.
26. LA Vigne &Owens & Hetrick. (2006). *Evaluation of Target's Safe City Initiative, Implementing Public-Private Paetnership to Adress Crime in Retail Settings.* U.S.: National Institute of Justice.
27. LifeTalk, Public Safety Communications E-Zine. (2012). Alcatel.Lucent.
28. Malik, A. R. (2003). Threats to Pakistan's Internal Security. http://www.defencejournal.com/2003-08/opi-e.htm.
29. Mujtaba, Y. (2014, 08 13). *http://advocacy.globalvoicesonline.org/2014/08/13/human-rights-experts-pakistan-could-become-a-police-state-under-protection-ordinance/.* Retrieved from http://advocacy.globalvoicesonline.org/2014/08/13/human-rights-experts-pakistan-could-become-a-police-state-under-protection-ordinance/: http://advocacy.globalvoicesonline.org/2014/08/13/human-rights-experts-pakistan-could-become-a-police-state-under-protection-ordinance/
30. Nadeem, D. A. (2002). *Pakistan: The Political Economy of Lawlessness*. Lahore: Oxford.
31. Nadeem, D. A. (2012, 07 27). Countering terrorism with National Consensus. *The Express Tribune* .
32. Nadeem, D. A. (2013, 03 22). Stop Paralysing the Police. *The Express Tribune* .
33. Nadeem, D. A. (2013, 11 04). Setting up an Antiterrorist Force - Strategy and Design. *The Express Tribune*
34. National Security Strategy of United States. (2010).
35. National Security Strategy of United States. (2013).
36. National Internal Security Policy of Pakistan. (2014)
37. Neill, A. (2010). Towards Cross-Border Security. *Occasional Paper.*

38. Olsen, M. G. (2012). Hearing before the House Committee on Homeland Security Understanding the Homeland Threat Landscape July 25, 2012, 01-11.

39. Pakistan's Internal Problems (all aspects). (2013, 04 22). Retrieved from: www. ppscpms.blogspot.com/2013/14/Pakistan-internal-problems-all-aspects.html.

40. Planning Urban Design And Management For Crime Prevention. (2006-2007). European Commission Directorate-General Justice, Freedom And Security.

41. Protection of Pakistan Ordinance promulgated to fight terrorists, criminal elements: Zafar. (2014). Retrieved from: http://www.app.com.pk/en_/index.php?option=com_content&task=view&id=246709 &Itemid=2

42. Reese, S. (2013). *Defining Homeland Security: Analysis and Congressional Consideration*. Congressional Research Service.

43. Riley, K. Jack, and Bruce Hoffman. (2010). *Domestic Terrorism: A National Assessment of State and Local Law Enforcement Preparedness, Santa Monica,* Calif.: RAND Corporation .

44. Safe City Moscow. (n.d.). www.axxonsoft.com.

45. Shamsuddin, S. B., & Hussin, N. A. (2013). Safe City Concept and Crime Prevention Through Environmental Design (CPTED) for Urban Sustainability in Malaysian Cities. *American Transactions on Engineering & Applied Sciences*, 223-245.

46. Tankel, S. (2011). A Pakistan-Based Terrorist Attack on the U.S. Homeland. *Council n Foreign Relations*, 01-09.

47. The National Security Strategy of the United Kingdom, Security in an Independent World. (2008). United Kingdom: Cabinet Office.

48. Towards a National Security Strategy for India. (2011). IDSA National Strategy Project (INSP).

49. U.S. Department of Justice, Office of Justice Programs, Global Justice Information Sharing Initiatives. (2006). *Fusion Center Guidelines Developing and Sharing Information and Intelligence in a New Era.*

50. U.S. – India Homeland Security Cooperation. (2013). Center for Strategic and International Studies.

51. Yarger, H. R., & Barber, F. G. (n.d.). The U.S. War College Methodology for Determining Interests and Levels of Intensity.

52. Zeb, R. (2006). Cross border Terrorism Issues Plaguing Pakistan-Afghanistan Relations. Islamabad: *Central Asia-Caucasus Institute Silk Road Studies Program*

CHAPTER NO. - 8

ECONOMIC SECURITY: CHALLENGES AND OPTIONS FOR PAKISTAN

The paper dwells into the Economic Challenges that Pakistan is facing today. The economic condition in Pakistan remains fragile as it pertains to balance of payment, inflation and public debt all of which paints a glum picture. In this paper, it has been endeavored to identify the root causes that adversely impact the economic growth which appends flawed fiscal policies, improper management practices, political instability and grimed law and order situation. These are some critical enablers which seriously hampering Pakistan's growth. But further, the paper also highlights that Pakistan has shown some positive economic indicators of a healthy economy as well. To propel this growth, there is a dire need for strong political and economic institutions with adequate long-term focus towards analysis, formulation and implementation of sound economic policies. It is also stressed that the growing population of educated youth in Pakistan is to be strategically placed in state's economic activities to turn it into positive human capital. Lastly, the warrant for the democratic institutions with a strong focus on decentralsation of governance to avoid abuse of power and formatting a system of accountability is inevitable for augmenting the economy of Pakistan.

The state of Pakistan is confronting multiple challenges since its birth. These challenges are exhaled from a complex and rapidly changing internal and external environment. Most of the external challenges are related to its foreign policy and much emphasized defense policy. All the challenges have direct link to the issue of Pakistan's national security. Although the traditional national security parameters were only revolved around the defense policy and foreign relations but the new national security patterns are involved internal threats. (Rizvi, 2012,p.9). The internal threats can be sound as social, political and economic. Despite the fact, that the three realms have their own jurisdiction but one are influence the other. The security and well-being of a state depends on proper functioning of social, political and economic life.

Under the parameter of national security, this paper will deal the economic aspect.

- **Economic Challenges of Pakistan.**

The economic condition of Pakistan is fragile. It remained challenging from 1947. From the last 67 years, the economic environment is affected by various domestic and global factors, which not only pressurized the state from above but from below as well. Resultantly, the economy moved towards a substantial decline in all the areas. At present overall economic progress is slow. Inflation is nearer to double digit (Ahmed, 2014,P.I), there is deficit in balance of payment, the state usually facing low level of foreign reserves and increased public debt. For overhauling the economy, a strong psychological and physical policy is required, having ability reduce the risk of bank rusty and save the homeland, citizen and their way of life. Economic security is also referred to in terms of human security. If individuals have buying power and economic opportunities, it means the economic system has the capacity to give self-reliance. Economic issues in Pakistan are viewed in the following figures.

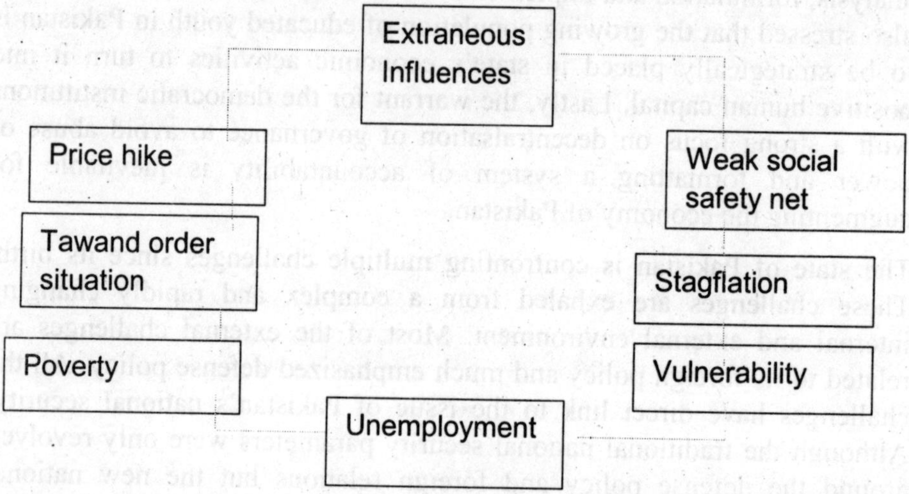

Source: Figure adopted from khan,M.E.(2013) Human security in Pakistan. Lahore BPH printer (p.83)

In the above figure some important issues are highlighted, creating economic insecurity in Pakistan. A part from these issues the economic challenges to Pakistan can be summarized into three areas.

- Policy Problems
- Management lapses
- Domestic environment

- **Policy Problems.**

Owing to the fact, that Pakistan is a developing nation and has less capacity to follow the standards of globalized economy as there are various flaws in its basic economic policy. Due to this ill policy making, following issues are influencing the economy negatively.

- **Fiscal Deficit**

Usually annual fiscal policy is made to achieve certain objectives for economic growth. These objectives include reduction of poverty, job opportunities and development of human and physical infrastructure unfortunately, due to regular fiscal deficit. These objectives are not achieved. The reasons behind are imbalance in revenue and expenditure and a frequent change in governments fiscal policy, security risks and power shortage at domestic level, while the global financial environment also did not a conducive role the global financial crisis which was erupted in late 2007, but its consequences can be seen in the developing economies obstructed the central bank's ability to manage the economy. All these domestic and global factors became a cause of deceleration in investment rate, unemployment and poverty and the basic aim of stabilizing economic growth is yet to be achieved.

1.1.2 Low export growth and excess of imports.

It is any unusual fact that 80% of Pakistan's imports in 2007 and 2008 were financed by its export earnings. Initially from 1960s import substitution industrialization policy was made (Noman, 1988.P.74) under this policy it was tried to set up local industry that would have the ability to produce those goods locally which were being imported earlier. Resultantly, with the collaboration of foreign power a large capital goods producing unit such as steel mill was set up in the first tenure of Pakistan People's Party. Lately, due to various regional and international issues, the exports of manufactured goods from Pakistan drastically reaches on the lowest ebb. Once after the East Pakistan crisis when Pakistan faced cessation of foreign aid as a punitive measure against its armed action in East Pakistan. Second, when Pakistan made its nuclear tests and third, after the incident of 9/11. The effects of 9/11 incident are still very significant. From the developed market, the demand for export became slow down especially from the Muslim countries. Pakistan was also included (Zaidi, 2008. P.6) although the state of Pakistan is continuously trying to explore new markets for tis exportable items but international response is not sufficiently favorable. In contrast to export items, the import of goods is increased substantially. In Pakistan, people usually prefer to use import items in their daily life which has widened the gap between export earnings and import expenses. According to the Pakistan economic survey (2011-12) the trade deficit expanded because the growth rate of imports·

1.1.3 Low domestic savings.

According to Dr. Ishrat Hussain, former governor of State Bank of Pakistan, "We consume 85 rupees and save on 15 rupees, out of one hundred". (Hussain, 2009) it means a very little amount is available for investment. According to him the required rate of saving should be 25% for only 06% annual growth. Other regional contenders, like India and China have much high rate of savings that is 25 and 50% subsequently.

1.1.4 Low percept Income

According to economic survey of Pakistan 2011-12, precipitate income of Pakistan was $1372. In 2013-14 the figures are not drastically changed (Pakistan economic survey 2011-12 P.ii). Precipitate Income has a strong relationship with the population's growth rate. Today the approximate population of Pakistan is 170 million so the precipitate income is too low to meet the demands.

1.1.5 Burden of External and internal debt

To facilitate the development process, governments used to borrow from domestic and external sources. An effective and efficient policy can adjust the borrowing in repayment and economic growth otherwise it can plague economic development. Growing debt burden domestic and foreign both has remained the biggest problem. From 1980s onwards Pakistan's economy is continuously struggling to equivalent debt to its GDP. During Afghan war from 1980s to 1990s and after the incident of 9/11, a huge amount of debt was written off and rescheduled in the back drop of Pakistan's support to USA in war against terrorism (Zaidi, 2008, p.8). The debt burden has been lifted for a short span. In the year 2004-05 highest growth rate has been observed. Exports crossed the figure US$ 17 billion for the first time.(Zaidi, 2004,p.) Unfortunately in the following years, the economy could not maintain the same pace.

Since the fiscal year 2007some substantial changes have been observed in Pakistan debt dynamics. Due to countless domestic issues, and unfavorable global economic environment. The debt position of Pakistan is influenced greatly. The fiscal and current account resulted in deficits. Fiscal deficit is a great challenge to Pakistan's economy Because of this situation the economy of Pakistan is mainly relations international financial institutions (IFLs), on foreign countries / donors and on the remittances from overseas Pakistan (Rizvi,2012, p.188). Due to security situation and energy crisis, foreign direct investment (FDI) reaches at its lowest ebb. The situation is challenging on both the sides public debt position and external debt is in declining position. Following figure shows the external debt liabilities to Pakistan from 2008-2013

Figure external debt

Figure 1.1

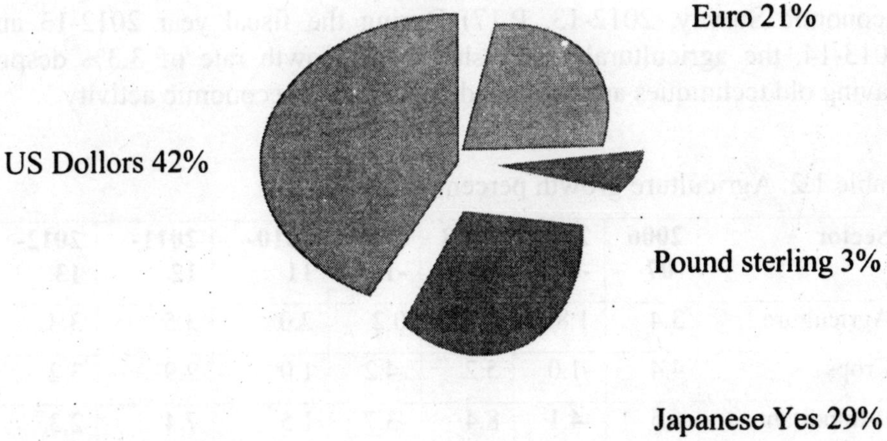

Table 1.1: Pakistan's Public External Debt Servicing

Year	Actual amount paid	Amount rolled over	Total
2008	3,182.6	1,200.0	4,382.6
2009	4,747.4	1,600.0	6,347.2
2010	4,607.0	1,723.0	6,330.0

2011	3,947.7	1,488.0	5,435.7
2012	4,507.7	1,543.0	6,050.7
2013	4,602.1	700.00	5,302.1

Source: state bank of Pakistan and Debt policy coordination office staff calculations

1.1.6 Over dependence on agriculture and small industrial sector

Pakistan's economy is highly dependent on agriculture sector, and it is central to economic growth and development. Main reason is that about 61% of population is the resident of villages. Where agriculture is the main source of income. Unfortunately agriculture sector did not adopt the new techniques. In most of the areas of the four provinces of Pakistan old and traditional methods of cultivation and farming are still in practice. In contributes 21.4% to GDP and it employed 45% of labor force. (Pakistan Economic Survey, 2012-13, P.17) During the fiscal year 2012-13 and 2013-14, the agricultural sector showed a growth rate of 3.3% despite having old techniques and remained central to the economic activity

Table 1.2: Agriculture growth percentage

Sector	2006-07	2007-08	2008-09	2009-10	2010-11	2011-12	2012-13
Agriculture	3.4	1.8	3.5	0.2	2.0	3.5	3.3
Crops	4.4	-1.0	5.2	-4.2	1.0	2.9	3.2
i) Important Crops	6.5	-4.1	8.4	-3.7	1.5	7.4	2.3
ii) Other Crops	2.1	6.0	0.5	-7.2	2.3	-7.7	6.7
iii) Cotton Ginning	-0.8	-7.0	1.3	7.3	-8.5	13.8	-2.9
Livestock	2.8	3.6	2.2	3.8	3.4	3.9	3.7
Forestry	2.7	8.9	2.6	-0.1	4.8	1.7	0.1
Fishing	0.8	8.5	2.6	1.4	-15.2	3.8	0.7

One of the reasons of this over dependence is the small industrial sector as compared to the other developing states. Although Pakistan had make a structural shift towards industrialization or manufacturing because it is also a main source of economic development, but could not replace the formal. In the last fiscal year 2013-14 the target was 6.1% but it had contributed only 3.1% share of GDP, which is negative growth rate. Overall its share in the economy is 13.2% of GDP and 13.8% of total employed labour force. There are various reasons behind this poor performance including the strategically and technical weaknesses but at present energy crisis and deteriorated law and order situation has the main role. Due to the reason, it is impossible for the exporter to meet their targets so exports are confined to raw material only. And we have to pay much more on finished value added products. In 2012 European Union announced a duty waiver on 75 products and GSP Plus status in also operationalized in 2014 that is a positive sign for the development of industrial sector in Pakistan.

Managerial Lapses.

Proper management of economy can ensure economic security. Apart from policy problems, there are various managerial issues, hampering Pakistan's economy.

1.2.1 Mismanagement in Collection of Taxes

Raising public revenue is a serious issue in Pakistan. Due to consistent failure in the generation and improvement of revenue the economy of Pakistan is heavily relying on borrowing (Chaudhary and Hamid, 2001, p.25) the government used to adopt cosmetic measures and adhoc policies to improve taxation, but always failed to get desired result. For economic growth the importance of tax revenue cannot be ignored. In Pakistan government is failed to develop a strong policy for the collection of taxes that is why only one percent people in Pakistan are taxpayer. The social services structure is also dependent on taxation. Almost half of the population is derived from basic necessities of life. They are facing poor health and sanitation facilities, even basic education is not in their reach. The tax performance is consistently being poor day by day. The most important institute for tax collection is the federal board of revenue (FBR). The tax collecting machinery including FBR are not doing through to widen the tax net, and allowing tax evasion and tax exemption at a massive scale. According to a report from Express Tribune, "the land owning class have been evading taxes to the tune of over $1.2 billion a year (Haafiz, 2012)

Apart from the exemption of agriculture sector, there are number of statutory regulatory order (SROs) to give tax exemption to various kinds. According to custom's intelligence report these SROs are used negatively. It is also reported that about 4,000 manufacturing units who had received tax exemption for importing raw material, did not exist (Farooq and Zaidi, 2013, p.42). According to Pakistan Economic Survey 2012-13 the national exchequer has lost 119,706.3 billion rupees in the fiscal year alone because of just 15SROs issued between 2004 and 2009 (p.59).

It is also noticed that there are specific internal conflicts among the officers of FBR who belonged to two groups (customs group and income tax groups) is also a source of inefficiently. The problems in the structure and working of FBR allows the institution to perform badly. The net impact is the loss of a huge amount of revenue. Government should focus on the issue to formulate an equitable national tax policy.

1.2.2 General Poverty and low living standards.

Generally, poverty is considered as the base of economic insecurity. According to United Nations reports, poverty is not a missing feature of the landscape of the developed economies. The ratio varies in developed and developing world. The prescribed yard stick to measure poverty in the world remained changing time by time. Once I was a dollar-a-day per person valued at the 1985 prices then revised to 1.08 dollar per day based on the 1993 price index after that it was replaced as 1.25 dollar per day based on 2005 price index. At present, in most of the countries US dollar 01, is considered a basic standard to measure poverty and around 1.4 billion people are living in (United Nations, 2010, p.1). in Pakistan's economy, poverty is also a major stumbling block. The planning commission of Pakistan presented national poverty line in 2000-01, on the basis of certain human development indicators. The Physical need of an adult person per day are 2350 calories per day at Rs. 748.57 per capita per month at the price of 2000-01. (Khan 2013, p.85). In Pakistan there are several reasons of poverty on one inside it in the result of the poor and ineffective policies of the government and on the other side people by themselves are also responsible for this phenomenon. The government of Pakistan is completely failed to create a balance between its resources and population. The population's growth rate is at the highest end. Although the state institutions are planning and working for this cause but there are lots of irritants in the way of executing and implementation of this policy. Political Parties and religious groups can help government by using their influence and by socializing people about this issue but they are not doing this job. Resultantly poverty is increasing day by day Government is failed to provide the shelter, schools for basic education, Hospitals for health issues, job opportunities for educated people. Health environment and security of life is also not possible. For the reduction of poverty state and society both has to make a strong collaboration to determine their future needs. Government has to control rising prices of food commodities. Social security networks and social protection of public and private sector should be made possible. Economic opportunities should be increased on equal footings so that every strata of society could explore them.

1.2.3 Un employment and brain drain

Unstable and mismanaged economic system is the major reason of unemployment. In Pakistan the ratio of unemployment is 7.4%. One of the major reasons behind this increased phenomenon is the inappropriate education policy of the government which is not objective oriented. New educational institutes including professional and technical are opened and have many branches but these are not properly linked with the existing industries except few. A large number of educators are passing out every year and they have no space in the market. This is a great lapse of the government policies that they cannot bridge between the market requirement and educational policy. Following table can highlight the rate of unemployment year by year

Table 2.1: Unemployment rate in Pakistan 2005-2011

Unemployment rate (%) population	2005-06	2006-07	2007-08	2008-09	2009-10	2010-11
Male	5.2	4.2	4.0	4.2	4.1	4.8
Female	9.6	8.6	8.7	9.0	9.2	9.0
Total	6.1	5.1	5.0	5.2	5.3	5.7

Source: (khan, 2013, p. 88)

Due to week policies of the government, unemployment is increasing day by day that result s in brain drain. Educated people who do not find jobs are migrating to abroad for better future. The state of Pakistan is in need of skilled and educated learned people. Proper and in time policies can stop this brain drain.

1.2.4 Inappropriate Use of Natural resources.

In Pakistan, there is no shortage of natural resources, it is endowed with huge potential of natural resources including precious metals, industrial minerals rock salt and coals. The problem lies only in the limited exploration. Pakistan lacks modern management, adequate capital and appropriate technical knowhow in this field, which is the main reason of backwardness in this area. The government is required proper policy planning for mapping in different to have parts of the country. Most of the natural resources especially mineral deposits are concentrated in the province of Baluchistan. Huge reservoir of natural gas, gold, copper and petroleum are available in Baluchistan. Baluchistan was contributing nearly 56% to Pakistan's natural gas output till 1995, but gradually this ratio is declined and now it reach almost 5.81% of country total output . This is because of the deteriorated law and order situation and inappropriate policies of the government

1.2.5 Energy Crisis and water shortage

Persistence of energy and water shortages is an important issue, which has become a challenge for the stage of Pakistan. It is one of the important reasons of slowing down the economy. On rational ground energy crisis in Pakistan is seriously an outcome of mismanagement and vision less planning. Energy requirement are enhancing day by day but the generation capacity of existing power plants is not enough. The situation is getting worse because of inefficiency of Public utility companies, high distribution losses, nonpayment of electricity bills and the circular debts. Due to energy crisis, our industry is at high loss. They are unable to execute their orders in time. So a major portion of manufacturing industries is shifting to nearby stages especially in Bangladesh. On this issue serious concerns of the government is needed because new large storage recover are needed. Our dams are silting but not a single new large reservoir is constructed after Tarbella dam. Rain water loss is also a great issue because there is not adequate management and after every year or two due to excess of rainfall, floods further worsen the situation. Apart from this situation, international Oil prices are also fluctuated widely. Among all the vulnerable states, Pakistan has no exception. It remained under great stress because Pakistan use Oil in electricity production. The erratic situation has not only made major dents in the economy of Pakistan but also abraded the purchasing power of the poor strata of society. Statistical report of Pakistan Economic Survey on energy consumption by source brought into light that gas and Oil has largest share.

Figure 3.1: Share of Source in Energy Consumption

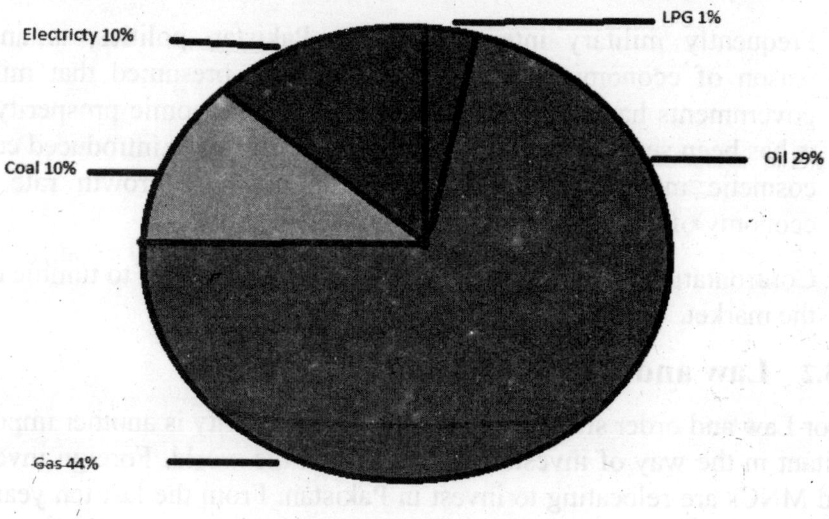

Domestic Issues

A well-functioning economy needs to have Political stability in the state. Unfortunately, the state of Pakistan has become a state in crisis (Cohen, 2012, p.xi) there are numerous issues, influencing directly the economy and causing economic growth slowdown.

1.3.1 Political Instability

Political instability is the major reason of deteriorated economic situation in Pakistan. Since its inception, political institutions are not working continuously. Due to the reason a coherent and integrated policy to manage national economy has remained a dream. Apart from the fact, that in that in the global economic environment market forces have become so strong and able to develop rules of the game but in Pakistan, government still has decisive authority, due to the reason. Pakistan's economic stability is dependent on its political stability there are various reasons of this instability which hampering the economic conditions some are as follows,

- Frequent use of article 58/2B off and on dissolved the representative governments. These government did not complete their political terms. Resultantly, economic policies are managed on adhocism, short terms planning did not contribute towards sustainable economic development.
- Frequently military intervention into Pakistan politics, is another reason of economic insecurity. Although is presumed that military governments have contributed a lot towards economic prosperity, but is has been seen clearly, that these government have introduced certain cosmetic measures, which showed an artificial growth rate. The economy of Pakistan is still facing its negative impact.
- Confrontational Politics is another reasons which used to tumble down the market.

1.3.2 Law and Order situation

Poor Law and order situation and increased insecurity is another important irritant in the way of investors from all over the world. Foreign investors and MNCs are relocating to invest in Pakistan. From the last ten years the ration of foreign investment decreased due to the negative image of Pakistan and growing concerns of foreignness towards their security. The internal disorder is increased since Pakistan's participation in war against terrorism. Instances of organized violence by different militant outfits have increased especially in the Province of KPK and Punjab.

1.3.3 Poor governance and implementation issues.

Bad governance is also a major reason behind economic instability I Pakistan. Government is badly failed in managing its education, health agriculture, industry, trade policy and other department. According to latest report of Transparency international. Pakistan has been ranked at 42nd number among the most corrupt nations of the world. Overall indicators of governance are also show negative trends (Mughal, 2014, p.41). According to world democracy audit scores and Ranking 2010, Pakistan is as follows.

Topics		
World democracy audit overall ranking	1-150	108
Political rights	1-7	6
Civil liberties	0-1	5
Freedom of press	0-100	95
Corruption	0-150	117

Source: http//www.worldaudit.org/countries/Pakistan/htm.

Weak and deteriorated political system with the problems of good governance automatically derail the economic system. Major reasons of Poor economic governance can be summarized as,

- Since long, government lacks expertise and capability to manage human resources.
- All the government had short term economic objective limited to managing fiscal affairs only.
- Inability of leadership to run the government posts especially its economy.
- Centralization of decision making
- Inappropriate policy of taxation and inability to enhance potential taxable areas.
- Excessive reliance on external funding without proper planning (Yusuf, 2012, p 271)

1.4 Poor Social Indicators

There is a strong link between social indicators and economic development. Social indicators include literacy rate fertility rate, infant's mortality ratio, access to basic services, etc. Among other Asian countries Pakistan is lagging way behind. The literacy rate in all the provinces of Pakistan is not as much as it is required. Pakistan is spending only 1.8% of GDP on education. Health condition is also very poor. Many people are fighting continuously against malnutrition. Average life expectancy in Pakistan is not more than 65 years. Infant mortality rate is enormously high, comparing to other South Asian countries. All these issues cumulatively create an environment which is not conducive for economic development.

2. Prospects for Economic Security

The above mentioned factors present a very bleak situation for economic development. Despite all these issues, the economy of Pakistan is considered a potential economy because it is growing. In agriculture, Pakistan is the third largest rice exporter of the world. Pakistan is one of the leading cotton producing countries. At this time industrial sector is gradually rising and about 40% of labour force is engaged in agriculture, while in the past the ratio was 70 to 75%. Pakistan has also placed on third position as for as the milk producing countries are concerned fisheries of Pakistan once again placed itself in the European Market and it has got the status of GSP plus. These are health indicators for a growing economy. Now the question arises how to overcome the challenges and enhance our potential. For this purpose state has to evolve a multi-pronged strategy which would be able to change the overall behavior and mindset, of the nation and to enhance the capacity of the institution to diagnose, suggest and then implement the plan properly.

Following are the key points in this regard.

2.1 Change in National attitude and mindset.

The most essential element for a strong strategy is to change the set attitude. It is too much cyrical that in Pakistan everything perceives negatively. This mindset should be changed and it is possible wen we can highlight the positive image. We have to rely on ourselves and should not wait for a savior. Everybody has to feel their responsibilities and should carry them. Unless we have a positive approach it will be difficult to progress.

New role of state and policy parameters.

In most of the developing countries still state is performing its central and traditional role meeting the demands from local regional and transnational level. In this age of globalization, where state has many competitors, holding central power and authority is difficult to maintain. In Pakistan state remains central to the wellbeing of its citizens and in the management of social and economic development. It means that now state is to call upon to act as linking pin among various processes include planning consultation, negotiation and decision making because of the involvement of diverse players, state and non-state actors at different levels of governance

A state with weak political and economic institutions cannot provide a stable and conducive environment to attract foreign investment of the promotion of economic growth and for fully integration with the world economy Pakistan needs to modernize its public sector and should enhance its capacity for policy analysis, policy formulation and policy implementation. For this purpose, it is necessary to build up

- Diagnostic capacity
- Strategic capacity for the design, the conduct, for monitoring and evaluation of program of reforms.
- Office bearer must hold deep knowledge, knowhow and range of skills in order to be able to design, implement and evaluate effective national policies.
- To conduct the consultation and negotiations at both national and sub-national level, in order to elicit, articulate, promote and defend the public interest.

There requires professionalism in public policy and development leadership skills and strategic planning capacity, program implementation in human resources, management and development.

Under this define new role, following policy parameters are necessary to opt for economic security.

Appropriate Policy for revenue generation or mobilization

In Pakistan an important reasons of fiscal deficit is low revenue collection. There is dire need to widen the tax net and improve tax collection strategy. For this purpose it is necessary to remove exemptions and concessions to various personalities. National saving rates should be increased. By tightening enforcement and compliance, tax revenue could be enhanced. This should also carry out proper audits and supervision of tax administration.

Focus on building human capital

Human capital is the actual power of state. Those countries who are proper and developed, did not compromise on human capital. It means that state will ensure to provide, better education facilities proper health care system and a conducive social and political environment to develop. Pakistan is in need to join hands with private sector, NGOs and various local groups to ensure that no child will left behind. Proper education which enhanced their ability to perform and provision of level playing field for all rich and poor will be helpful for the development of human capital.

Appropriate use of youth potential

In Pakistan the ratio of young men and women is going to rise. Most of the developed societies are facing the problem of aging population. To create balance these countries are giving attractions to the young educated and skilled people of Asian countries, Pakistan has no exemption. Due to low wages, and widely spread unemployment, the young and skilled people are going abroad. It is necessary to create opportunities for the youth, and to equip then properly. Inadequate employment opportunities could be caused social upheaval. Efforts towards the building up of this potential could be fruitful of the state have proper demographic record of each and every group in the state. Unfortunately, in Pakistan authentic demographic record is not available because of huge gaps in the census. After 1998, government did not issue even single senses report, which is an essential part of policy formulation

We are making estimates due to the reasons neither the resource allocation policy is successful nor its distribution policy. The census department should show their concern on this issue.

Security, stability and decentralization

Issues of governance, security and stability are hampering the economy of Pakistan. Strong institutions and well-functioning governing structures foster economic growth. Pakistan should improve security situation and institutional stability. This is possible when actual spirit of democracy should be realized that lies in the decentralization of power. Concentration of power causes abuse and decentralization promotes and enhances efficiency, accountability and responsiveness. Devolution of power to local or district government will create capacity building at lower level. Civil services will be more efficient and responsive to meet the needs of the commoner's. Under the Hierarchy of the government, local government will also be accountable before provincial government and provincial before the federal. Following the jurisdiction of the government and institutions will create a system of checks and balance to avoid unwanted penetration and interference.

Conclusion

After a long discussion about challenges facing Pakistan's economy and the new role of the state to overcome these challenges one can argue that national security of Pakistan can be strengthened if there is economic security. It is at the heart of national security policy.

References

1. Ahmad, K (2013). Pakistan's economic condition remains challenging-IMF. Pulse weekly, January 13, 2014. Islamabad. Retrieved from. Weeklypulse.org
2. A lecturer delivered to the participate of the command and staff course at Command and Staff College Quetta. On September 04, 2009
3. Cohen, P.S. (2012) The future of Pakistan, Lahore: vanguard book
4. Chaudhary, A.M and Hamid, A (2001) Resource Mobilization and Tax Elasticities in Pakistan. Pakistan Economic and Social review Vol. XXXIX, No. 1 summer (p.p. 25-48)
5. Farooq, U and Zaidi, M(2013) Net Impact Monthly Herald, September (p.p. 40-44) Karachi.
6. Government of Pakistan, Finance Division (2011-12) Pakistan Economic Survey Islamabad: Economic Advisor wing.
7. Government of Pakistan, Fianance Division (2012-13) Pakistan Economic Survey. Islamabad Economic Advisor wing
8. Hafiz ,S (2012) Pakistan and its massive tax Problem. The Express Tribune, June, 7th, Lahore.
9. Mughal, R.M.(2014) Good governance in Pakistan: problems and proposed solution, International Journal of Modern Business issue of Global Market (IJMBIGM) Vol.2, issue 2, February 2014 (p.p. 40-58)
10. Noman, O (1988) The Political Economy of Pakistan. London: KPI Ltd.
11. Rizvi, H.A (2012) National Security Council-A debate on institution and process for decision making on security issues, civil-military relations. Islamabad: Pakistan Institution for Legislative Development and transparency.
12. Rizvi, H.A (2012) at the brink. (In) the future of Pakistan (Eds) Stephen.P.Cohen (p.p. 182-198) Lahore: vanguard books.
13. United Nations (2010) Rethinking poverty: Report on the world social situation 2010, department of economic and social affairs.
14. Yusuf, M (2012) Youth and the future (In) Stephen P.Cohen,(Eds). The future of Pakistan (p.p. 257-283) Lahore: vanguard.
15. Zaidi, A.S (2008) The political economy of military rule in Pakistan: the Musharraf Regime. ISAS working paper No. 31, retrieved from www.isas.nus.edu.sg
16. Zaidi, S.A (2004) Pakistan's economy after 9/11: will the end be different this time around? Occasional paper No. 6, 2004. Cambridge: centre for South Asian studies, University of Cambridge (p.p.)

CHAPTER NO. - 9

HUMAN SECURITY: A COUNTER-NARRATIVE TO RELIGIOUS EXTREMISM IN PAKISTAN

The paper highlights the need for security in terms of freedom from fear, coupled with freedom from wants which is essential for ensuring lasting global peace. The freedom from wants is referred to the basic rights of an individual towards leading a life of dignity and access to basic services like food, health, environment and education. Today, the human security has become the victim of growing religious extremism. In Pakistan, the sharp differences in normative values in terms of social, cultural and religious beliefs within different sections of its society perform as the bleeding fault lines. To counter the situation, Pakistan has waged a war against religious extremism and terrorism but a comprehensive understanding of underlying factors like social causes of human insecurity along with political and economic reasons have been overlooked. The instable political affairs can be attributed to the overbearing role of the military where as the government can be ascribed to the economic deterioration which impaired individual's sociological-psychological character. The fragile government institutions in the country paved way for the "black law" or Frontier Crime Regulation (FCR) law which is another major source of breeding extremism as this law disables jurisdiction of federal political and judicial agencies. The social security impressions include health, education, food and human rights which are pertinent to address the human security problem in Pakistan.

Conceptual Framework:

Human Security

Security has been the basic feature of primitive societies. The advent of modern society has brought significant changes to the notion and perceptions of security. This was primarily due to three factors: Increased technological developments, complex interactions of societies within and without and Meta knowledge produced on issues that affect human life. Until the mid-20th century, as the state was the focus of most analysts in understanding the people in that state. This macro analysis had more mechanical approach. During the last quarter of 20th century analyst started focusing on human development apart from developments by the state. Concept of Human Security is derived primarily from the Human Development. The foundation of Human security is said to be the United Nations Human Development report of 1994. However, the idea of human security has evolved years earlier.

At global level, the concept of human security was introduced after the two Global Wars that killed hundreds of thousands people in almost all parts of the world. In June 1945, US Secretary of State, in his official report to the US government states:

"The battle of peace has to be fought on two fronts. The first is the security front where victory spells freedom from fear. The second is the economic and social front where victory means freedom from want. Only victory on both fronts can assure the world of an enduring peace.... No Provisions that can be written into the Charter will enable the Security Council to make the world secure from war if men and women have no security in their homes and their jobs." (UNDP, 1994)

The statement above represents the understanding of dangers to humanity. Traditionally this danger was considered to be emanating from violence as has been referred to as 'fear'. Freedom from this 'fear' was considered one aspect of security. However a new dimension in post-World War Two about security was developed and named as 'want' related to social and economic front. This clearly implies the freedom from basic necessities of life specially poverty. So freedom from want was referred to as a new challenge to the security of an individual. However, with the advent of Cold War and obsession of global political leadership with traditional security against opposite camps, the idea could not flourish much. During this period, there is little or no academic debate on the threats to individuals from means other than violence. However, the notion of development gradually changed with increase frustration of traditional indicators felt to be inadequate to analyze and understand broad concept of security. Economic growth was considered to be too mechanical and a gap between growth and societal development led some economists to challenge it. To them, inclusion of human development in overall economic growth will complete the economic foundation of a society. Ultimate of economic growth and for that purpose the state, is to give a quality life to its people that is impossible without providing 'freedom from fear' as well as 'freedom from want'.

United Nations Development Program's Human Development Report of 1994 is the pioneering document highlighting the concept and its significance. The report has tried to alter the traditional concept of Security and emphasized that focus of UN and states' efforts should be human security at micro level rather than at traditional security through arms. The focus of the report on individual is highlighted as:

"In the final analysis, human security is a child who did not die, a disease that did not spread, a job that was not cut, an ethnic tension that did not explode in violence, a dissident who was not silenced. Human security is not a concern with weapons – it is a concern with human life and dignity."(UNDP, 1994)

The report goes further in explaining four basic characteristics of the concept 'Human Security'. First, Human Security is universal. People around the world require human security irrespective of the state they live in. Second, the components of human security are interdependent. Globalization has made national borders largely irrelevant or at the minimum less restrictive. Human insecurity issues like famine, terrorism, environmental issues transcend national boundaries. Thirdly, prevention rather than later intervention is cost effective for issues of human security. Finally, it is concerned with people and thus people centered approach. It is concerned with everything that concerns humans from physical needs to values and freedoms (UNDP, 1994). The report then highlights seven threats to Human Security namely:

- Economic
- Food
- Health
- Environmental
- Personal
- Community
- Political (UNDP, 1994)

Underlying argument by researcher is that improvement in above stated threats to human security, in case of Pakistan, is key in any attempt to eradicate religious extremism. Before moving into the debate, the second concept i-e Extremism is discussed in the lines below.

Religious Extremism:

Extremism is derived from word "Extreme" meaning any opinion favoring immoderate uncompromising attitude, policy or belief. However, there are various definitions of extremism depending upon the discipline and particular form of extremism. An eminent psychologist Dr. Aaron T. Beck define extremism as " It is a perspective or viewpoint that demonize those with opposing viewpoints, using all-or-none thinking to characterize people as either "fully for" or "fully against" the group's views, and identifying certain people outside the group as having less innate value on a basic human level" (http://zorach.wordpress.com/2010/10/12/definition-of-extremism-identifying-handling-extremist-views/) It is pertinent to mention here that terms like extremism, fundamentalism and radicalism are often used interchangeably; however, there is fundamental difference between them in many different ways and with different unspoken connotations.

The term extremism found its way into political language in many different ways without being labeled as a separate terminological category. In the 19th century the term extremism had become customary to parallel the extremes of the political spectrum from the vantage point of constitutionalism and besides having obvious difference, to work out the analogies and structural similarities. The pattern continued even in the early 20th century political discourse. In our contemporary times, extremism is multi-dimensional issue. It can be Political, Religious, ideological and can be based on one's value system.

Human Security and Extremism:

There is dearth of literature on possible and plausible links between human insecurity and rise of extremism. There is, nonetheless, no dearth of literature highlighting the relationship between extremism and one or the other components of human security mentioned earlier. As recently as 2011, there were various studies that tried to de-link poverty, illiteracy etc. from extremism. However, most of these studies are quantitative and ignore the normative factors in the society that play a critical role in shaping the belief system and viewpoint of an individual in a complex society like Pakistan. Besides, the sharp differences in normative values within Pakistani society make a survey not applicable to the general society. The social, cultural and religious beliefs vary significantly in Pakistani society at different places. Religious extremism is almost non-existent in rural Sind and South of Baluchistan; however, this is not true in North West of Pakistan as well as South Punjab. The argument here is that qualitative and sociological analysis of the relationship between the two variables best help in understanding the causal relationship between them.

Most of the studies have by far, focused on religious (ideological), geo-political and sociological factors that play a role in growth of religious extremism in Pakistan. Human Security factor has so far been ignored by most scholars and especially by the government in her efforts to Counter Extremism.

Human Security in Pakistan

Background:

Pakistan is fighting a war within to curb violent extremism with traditional and military means ignoring the social and human security aspect. The religious extremism has crept in the society for last few decades and a culture of insecurity developed. Terrorism has not been born in Pakistan after the country joined the Global War on terrorism led by United States as frontline state. Rather this was triggering factor only preceded by a decade long support of religious zealots in her proxy war to counter external threats to the country. Pakistan inherited the threat of Indian aggression since Indian occupation of Kashmir in 1948. With balance of power tilting in favor of India, Pakistan had no other option but to use non-traditional means to deal with this threat. This was further added by tough stance of Afghanistan towards Pakistan and support by India of Afghan position on Pakistan. Joining Western capitalist block was the only option for Pakistan to balance the Indian threat and stabilize its defense which nurtured extremism exploiting socio-economic vulnerabilities.

Religion became the part of politics in Pakistan during cold war era when it joined the Capitalist bloc and propaganda began against Communism with US funding. This propaganda campaign initiated an era of religious politics and use of religion to exploit and achieve political ends. Pakistan was created on a religious ideology but the politics remained secular following the creation of this new nation-state. This is evident from the fact that the first Law minister of the newly born state was a schedule cast Hindu Jogandar Nath Mandal. United States found an opportunity to fund religious forces to fight Communism. Since then, the varied interpretations of religion had been used which played a vital role in dividing the nation. This gave birth to religious political trends and use of religion in managing external relations and internal politics. US proxy war with Soviet Union after she invaded Afghanistan, which proved the final war for United

States to fight Communism, was heavily dependent on religious indoctrination of Jehad against the infidels.

Pakistan joined this war to achieve multifaceted benefits. To provide force to this war, Zia regime again resorted to exploit religious sentiments to move the whole society in the favor of this war. This was the decisive era of nourishing extremism in Pakistan. Zia's Islamization affected whole social fabric deteriorating the tolerant ambiance. Pakistan's judicial, religious, political and education discourse transformed to radical and extremist discourse. The successful end of this episode brought extremism in to Pakistani society in the form of sectarianism. The extremist elements used this extremist and radical ideology as their refuge and survived in this guise. Pakistan joined War on Terror which was the immediate factor triggering the volatile wave of terrorism, militancy and sectarianism.

The support base to extremism was provided by the socio-economic worsened areas particularly in FATA and Southern Punjab. The religious seminaries developed in these areas during Soviet-Afghan war nurtured due to social inclination in less developed or not developed areas. The terrain of FATA gave the opportunity to hide at first place. It was further exacerbated by the fact that these two areas are least developed socially and economically. The have lowest level of educational, social, political and economic development. Terrorism and extremism have deep rooted relationship with socio-economic factors. Overall about 40% population lives below poverty line. This level further goes down to about 45% in least developed areas like FATA. These factors thus, enhance the volatility of extremism in Pakistan.

Political Security

Political Security is yet another major component of Human Security. Provision of fundamental human rights to the citizen is as important as other aspects of human life. Some of the key rights include right to life, right to justice (social and legal), protection from oppression & arbitrary persecution and freedom of expression. These rights and freedoms form the key to stable, peaceful and progressive societies.

Pakistan has a long history of military interventions in the politics of Pakistan. Military has ruled Pakistan more than half of its time since its birth. While most of the time civilian government ruled was primarily due to backing of Pakistan military. True democratic values could not nourish in Pakistani state and society. The direct military rule witnessed significant adverse impacts on the state of human rights especially against those who opposed their dictatorial rules. The military dictatorships are responsible for abrogating the first two constitutions and suspending thrice the third and current constitution. All these actions limited the political freedoms and have adversely affected the political security of the citizens of Pakistan. The arbitrary decisions and policies (shifts) by military regimes alienated many segments of society. People disown the decisions of dictators as they lack the local interests. This gap between ordinary people and state is exploited by certain elements to their own benefit. In Pakistan, this gap has primarily been exploited by conservative religious clergy to augment their own power locally.

Failure of different governments to develop the local government system is another political factor that kept people away from the policies of government affecting them. Local governments are considered primaries of any democratic setup. They help people own the political setup and this gives popular legitimacy to political system. The absence, on the other hand, increases gap between rulers and ruled, a gap often exploited by those with their own vested interests.

Finally, the most important issue breading extremism and terrorism is what is known as 'Black Law' i-e Frontier Crimes Regulation (FCR), a special law for Federally Administered Tribal Areas of Pakistan. FCR was established by British Colonial government in 1901 as special law to regulate the tribal affairs. Under that law, FATA remains a distinct part of Pakistan where no courts have any jurisdiction including the highest court i-e Supreme Court of Pakistan. It might be a unique feature that elected representatives form FATA cannot legislate on matters related to FATA. FCR allows the law enforcement to impose heavy fines to the tribes for not cooperating and in extreme case even the property of the entire tribe can be destroyed including their houses and this cannot be challenged in any court of law. One another ironic part is no political party can work in FATA region.

Economic Security:

Economic Security is the most critical of all the factors in Human Security in its correlation with extremism in the context of Pakistan. Although, the UNDP's Human Development Report of 1994, clearly suggests that incidence of poverty leads to social unrest, many studies suggested that poverty has no link with rise of extremism in Pakistan. The fundamental problem with these studies was that they were quantitative surveys linking these two variables. This methodology has two fundamental flaws in the context of Pakistan. First, these surveys ignore the sociological factors in shaping the respondents' answers. Extremism and like-wise phenomenon are highly sensitive in Pakistani society and more likely not given their true opinion. Secondly, there are some specific issues that are important and were ignored in some of these surveys. For example, in some regions of Pakistan, sending a child to a Madrassah is not indication of an extremist ideology, it might be, and most of times it is, out of economic and cultural compulsions. Therefore, a respondent would never conceive the idea of sending his kid to a religious seminary as a part of extremist idea. Therefore, we have to discuss the linkages between poverty and unemployment with extremism from a sociological-psychological framework. In Pakistan, there are few pocket territories that are considered pool of religious extremists' e-g FATA, Southern and western districts of KPK and South Punjab. Interestingly, in SDPI's report in 2012, these areas have highest incidence of poverty in their respective provinces. Following table shows the incidence of poverty in KPK:

Table: Incidence of poverty amongst the Punjab districts

Rank	Extreme poor districts		Rank	Least poor districts	
	Districts	Headcount Ratio		Districts	Headcount Ratio
1.	Kohistan	89	1	Haripur	11
2.	Shangla	63	2	Abbottabad	18
3.	Upper Dir	55	3	Nowshera	18
4.	Batagram	50	4	Chitral	22
5.	Bonair	50	5	Swabi	22
6.	D.I. Khan	41	6	Peshawar	24
7.	Swat	39	7	Charsada	24

8.	Tank	36	8	Bannu	25
9.	Malakan D	35	9	Kohat	26
10.	Lower Dir	33	10	Mansehra	26

Source: Naveed, Arif & Ali, Nazim. 2012. Cluster Deprivation: District Profile of Poverty in Pakistan.

The table clearly indicates the districts with highest number of poor people in different districts of KPK. The interesting point here is that the districts with highest number of poor people are most plagued by extremism and terrorism in KPK.

Table: Incidence of poverty amongst the Punjab districts

Rank	Extreme poor districts		Rank	Least poor districts	
	Districts	Headcount Ratio		Districts	Headcount Ratio
1.	Rajanpur	44	1	Jhelum	3
2.	Muzaffargarh	40	2	Gujrat	4
3.	D.G. Khan	36	3	Chakwal	5
4.	Bahawalpur	33	4	Mandi Bahuddin	6
5.	Layyah	31	5	Gujranwala	7
6.	Lodhran	31	6	Sialkot	7
7.	Pakpattan	29	7	Rawalpindi	8
8.	Multan	28	8	Faisalabad	9
9.	Khanewal	28	9	T.T. Singh	10
10.	Bhakkar	28	10	Attock	11

Source: Naveed, Arif & Ali, Nazim. 2012. Cluster Deprivation: District Profile of Poverty in Pakistan.

Various studies suggest the growing extremism in the southern districts of Punjab, commonly known as South Punjab. It is also believed that the Punjabi Taliban, an affiliate of Tehreek e Taliban Pakistan, is based in South Punjab. The above table indicates the prevalence of poverty in districts of Punjab. This table clearly mentions that all top 10 districts with highest proportion of poor people are from South Punjab.

Fig. District wise incidence of poverty (headcount ratio)

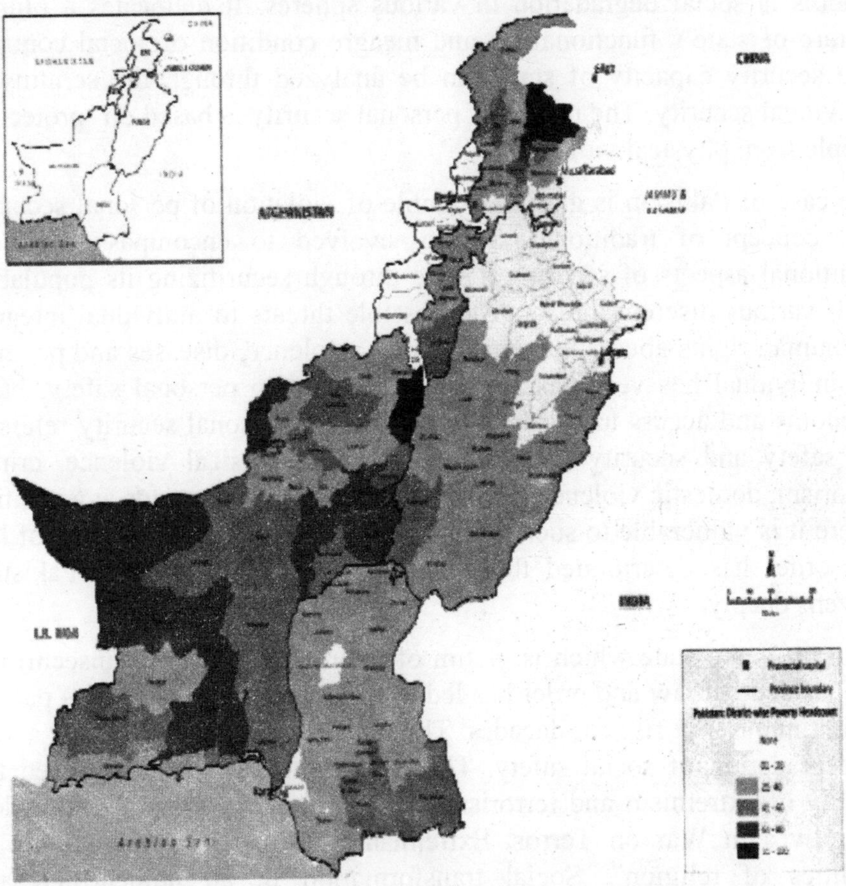

Source: Naveed, Arif & Ali, Nazim. 2012. Cluster Deprivation: District Profile of Poverty in Pakistan.

The map above indicates the concentration of poverty on the western half of Pakistani state. However, it is no coincidence that whole of this part is plagued with extremist violence.

Personal security:

Human security stands upon the very basic notion of securing the individuals fundamentally and ensuring their basic rights. The two fundamental aspects of securing individuals are health and education. The primary level is the protection of masses from vulnerabilities caused by lack of health and education facilities. The outcome of state's in capabilities to ensure individual health and satisfactory enrolments at schools is social degradation in various spheres. It delineates a blurred picture of state's functionalism and meagre condition of social contract. The security capacity of state can be analyzed through the scrutiny of individual security. The notion of personal security is based on 'protecting people from physical violence'

The case of Pakistan is glaring example of violation of personal security. The concept of traditional security evolved to encompass the non-traditional aspects of securing a state through securitizing its population from various discrete threats. The possible threats to individual integrity are human rights abuses, natural disasters, violence, diseases and poverty. An individual has very basic and certain right to personal safety, basic freedoms and access to sustainable prosperity. Personal security refers to the safety and security of individuals from "physical violence, crime, terrorism, domestic violence and child labor". Pakistan stands at a position where it is vulnerable to such insecurities immensely. The absence of law and order has deteriorated the personal security structure of Pakistani citizens deeply.

Pakistan is the state which is victim of internal and external insecurities. The absence of law and order has led to increased crime rate. This pattern is continuously at rise for decades. The menace of terrorism further added to degradation of social safety. This is the only state which faced the fatality of extremism and terrorism. Around fifty thousand civilians lost their lives in War on Terror. Extremism emerged with the culture of "politics of religion". Social transformation in an unidentified and undetermined direction converted the society in to ambivalent community. The external threats joined with internal insecurities and on the whole social fabric fell prey to extremist patterns. Currently, in Pakistan the individuals are victims to this pattern of personal insecurity.

Food security:

Pakistan is the agrarian state and major part of its economy is dependent on agriculture. Still Pakistan is unable to feed its population. The rapid increase in population is the major factor paralyzing the state to provide food to all its citizens. The malnutrition stemming from food scarcity is cause of various diseases and maternal and child health hazard. Food scarcity is the issue of world. Around one billion people remain hungry at night all over the world. Pakistan is among the states hit by severe food insecurity. According to Pakistan's National Nutrition Survey 2011, 60 percent of Pakistan's total population is facing food insecurity. Among them 80 million people are either hungry or victim of severe shortage of food. Women and Children are badly hit by this menace. About 50 percent of women and children are malnourished.

The issue of food security is more of accessibility than production volume. A country which is itself producing food can feed up to 2,600 calories per individual per day there about half of citizens are facing acute food scarcity. The reason is state's incapacity to distribute the resources according to the need of citizens. The average consumption is about only 1,600 calories. The gap shows that the problem lies in the governance and management of resources. Accessibility of food resources also limits due to inflation and less purchasing power of people. The environmental hazards also limit the capacity of a state to produce the food. Deforestation, lack of research and development, smuggling and hoarding are few reasons to reduce food production.

The implementation of a secure food policy is hurdled by gaps in legislative and implementation process. The increasing globalization had far reaching impact on the food sectors of various nations. The most affected counties are certainly from developing South. Pakistan is among those countries. The backwardness in technology and non- introduction of new methods in agriculture hinder the growth in food sector. Pakistan remained unable to regulate its food demand and supply. The outcome of such ill planning is delineated in the current food crisis badly hitting the nation. The recent calamaties in the form of floods in 2010, 2011 and 2012 devastated the agriculture. Pakistan was not ready cope with such situation which was visibly the failure of administration and governanace.

Food Absorption in Pakistan

Map 4.4

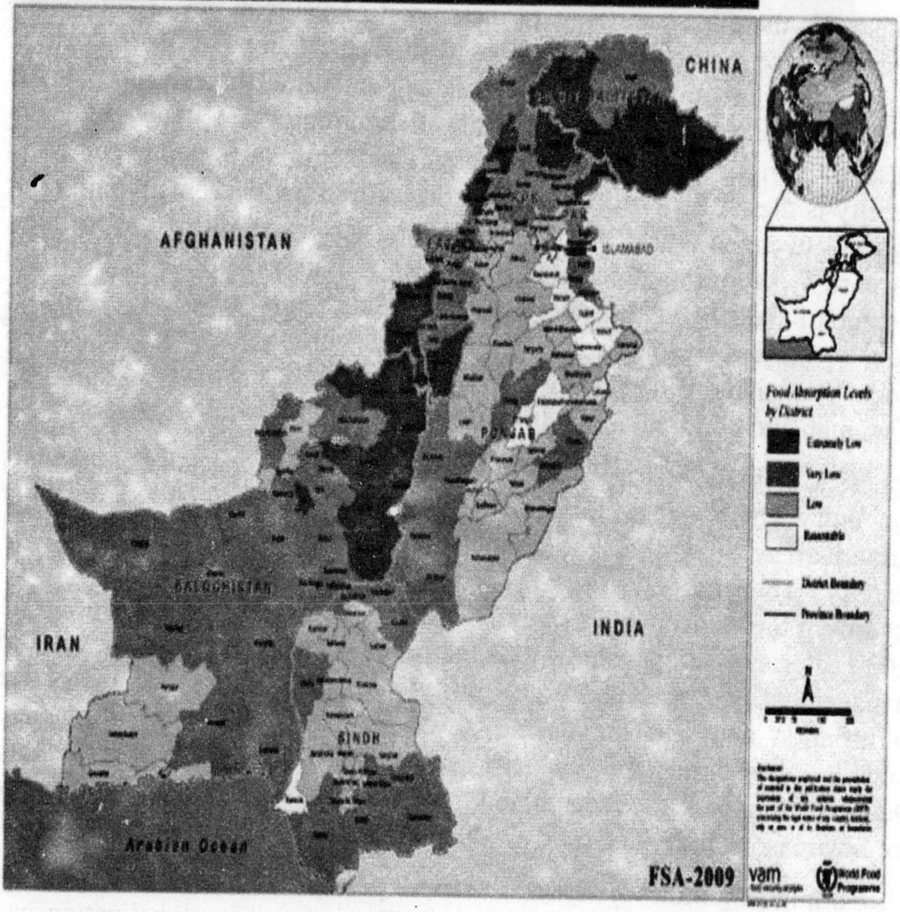

Source:
https://www.google.com.pk/search?q=food+security+in+Pakistan/images &tbm-isch&tbo=u&source=univ&sa=X&ei=Ov9eU5eyle6X0QWGzoDIBQ&v

Health Security:

Human security cannot be considered completely achieved until the physical health of individuals is ensured. Health of individuals provide with positive externalities to ensure social wellbeing. The individual health is essential for national security. Vulnerability to diseases, child mortality and maternal health are some key issues of health insecurity. The developed countries have achieved satisfactory level of health security.

The state responsibility to warrant this basic requirement to its citizens cannot be ignored. The dilemma of south in health sector is immense. The individual security in south is at stake which increases manifold in health sector. Pakistan is facing this acute problem for decades. Poor governance in health sector has caused massive insecurity all over the country which augments in less developed and rural areas. The urban population has other kinds of problems as over saturation in public health centers. The low budget allocation for health sector, mismanagement of resources and bad governance are the factors giving birth and escalating this problem. The World Health Organization constitution 1948, defines health as:

"Complete social, physical and mental well-being and not merely the absence of disease and infirmity"

The major chunk of population in Pakistan is still residing in rural areas and the health policies for rural areas are regressive. There is visible inequality of resource distribution in rural areas. Mother and Child health are worst affected by unequal and inadequate health facilities. The measures to ensure health security are also not adequately preventive. The variable trend of public spending on health is putting burden on state. The resource distribution is affected by social groups and impact of religion, caste, region and cultural background. The literacy rate and socio-economic realities also play their part in health improvement. Pakistan is signatory to Un Millennium Declaration 2000-2015. It had 18 targets and 48 indicators for health sector out of which Pakistan adopted 16 targets and 37 indicators. The primary emphasis was on child health, maternal health and prevention of HIV/AIDS, malaria etc. Maternal Deaths in South Asia reduced but less than Eastern Asia and northern Africa. This reduced by 69% in Eastern Asia, 66% in Northern Africa and 64% in Southern Asia.

The allocation of resources is a key factor in Pakistan in the development of health sector improvement. Insufficient resource allocation and then inadequate distribution increases vulnerability of individuals and social degradation. The total volume of budget of Pakistan 2013-2014 was Rs. 3,985 billion which increased by 24.4% than the budget of 2012-2013. The allocation to health sector was merely Rs. 9,863 million. This amount increased from Rs. 7,893 million in the budget of 2012-2013. This meagre volume of amount as compared to budget in total and ill distributed resources are worsening the health sector performance.

COUNTRYWIDE

Total budget of 2013-14 and its expenditure

	Rs in billion			As % of total budget		
Federation and units	Current	Development	Total	Current	Development	Total
Federal	59	21	80	74	26	2
Khyber-Pukhtunkhwa	59	14	73	81	19	21
Punjab	150	32	182	82	18	15
Sindh	121	14	134	90	10	22
Balochistan	25	11	35	70	30	21
Total	413	91	504	82	18	8

Source:http://tribune.com.pk/story/575729/budget-2013-14-despite-higher-allocations-education-still-not-a-priority/

Education security:

The developed countries of the world have high literacy rates and educational facilities as hallmark of their development. The knowledge and education are driving forces of any society. Socio-Economic development is also based on educational development. Economic security is ensured through educated masses. The world has started realizing the importance of education and socio-cultural development of community. Even the developing and the least developed countries are emphasizing on the importance and development of educational development. The increasing militancy and growing violence in South of the world has posed the world to curb this problem through education. Sri Lanka is glaring example of this problem. It enjoys a literacy rate of about 91.2% but it remained under the affect of conflict for decades.

Pakistan presents the example which is completely inverse. Pakistan's literacy rate is decreasing day by day. It is among the countries having least rate of literacy. According to the last conducted census in 1998, the government of Pakistan defined the literate person as one who can read newspapaer and write simple letter. But the government is devoid of measuring this through any mechanism and infrastructure. In 2009, Pakistan's literacy rate was reported as 55 percent by United Nations Educational, Scientific and Cultural Organization. In 2010, it was ranked 113th among the list of 120 member states of UNESCO's Education for All index.

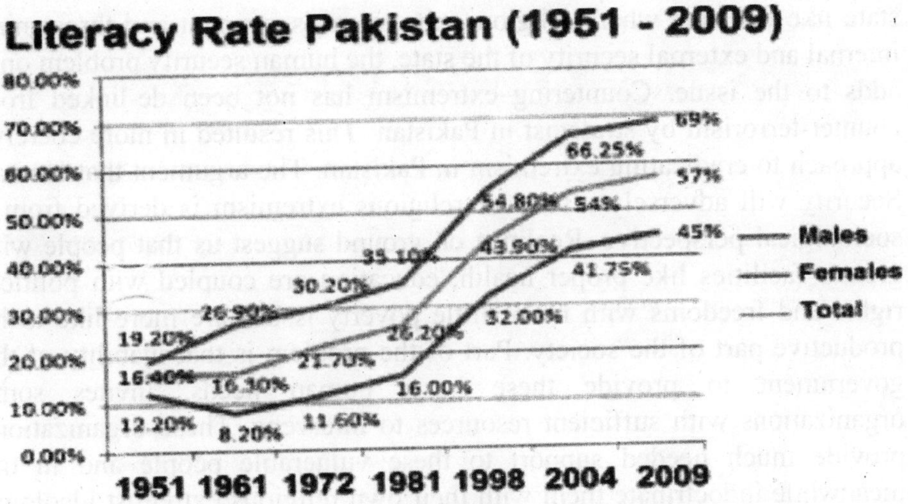

Source:
https://www.google.com.pk/search?q=food+security+in+Pakistan/images

Pakistan is under severe wave of militancy since it joined the War on terrorism after 9/11. This spell of violence badly hit the western border areas of Pakistan along Afghan border. The terrain of this region easily supported the militant hideouts. This support base for extremists had impact on education sector in Pakistan. Taliban in FATA destroyed various schools as they deemed it contradictory to their ideology. Around 600,000 children in Khyber Pakhtunkhwa missed one or two years of school due to militancy. The deficiency of teachers is another pressing issue. Teachers are not available in these militancy hit areas.

CONCLUSION

Human security has developed over the last two decades from a mere idea to a reality. As the UNDP's Human Development Report of 1994 predicted, most of the clashes taking place around the world are within the states rather than between or among states. Besides, threats to an individual are multi-dimensional and not limited to the traditional insecurities of violence. The issues of health, education, poverty, political repression, increased religious-based discriminations & violence and environmental issues are threatening an individual's life more than the armed clashes. This is truer for the developing societies in Asia, Africa and Latin America having majority of global population. However, for a state like Pakistan where religious extremism is rampant and threatening internal and external security of the state, the human security problem only adds to the issue. Countering extremism has not been de-linked from counter-terrorism by strategist in Pakistan. This resulted in more coercive approach to eradicating extremism in Pakistan. The argument that Human Security will adversely affect the religious extremism is derived from a sociological perspective. Realities on ground suggest us that people with greater facilities like proper health, education are coupled with political rights and freedoms with no or little poverty issues are more like to be productive part of the society. Part of the problem is that inability of the government to provide these basic human needs, invites some organizations with sufficient resources to intervene. These organizations provide much needed support to these vulnerable people and in the meanwhile indoctrinate them with their own religious extremist ideology. However, if a member of society has ample opportunities to flourish his abilities, he is less likely, be part of any non-state's extremist agenda. Finally, human security is not going to eradicate extremism from society altogether but surely, it will drain the recruits for extremists to a greater extent. This is a social a reality in the case of Pakistan. Any effort to eradicate extremism from society will have to involve heavy investment in the Human Security sectors divided in seven categories mentioned earlier. The idea got the attention of the policy makers recently in the newly formulated National Internal Security Policy by the government of Pakistan.

References:

1. Akhtar, Shaheen. 2009. "Human Security and Extremism in Pakistan: Synergy and Conflict." Islamabad: Institute of Regional Studies.
2. Alkire, Sabian. 2003. "A Conceptual Framework for Human Security." Oxford: CRISE, University of Oxford.
3. Chen, Licncoln. 2005. *Human Insecurity in Global World*. New Delhi: Viva Books Private Limited.
4. CSS Analysis in Security Policy. 2011. "Human Security: Genesis, Debates, Trends." Zurich: Centre for Security Studies.
5. Constable, P. (2011). *Playing with Fire*. New York: Random House.
6. Gasper, Des. 2008. "The Idea of Human Security." Hague: Institute for Social Sciences.
7. Human security Unit, U. N. (2010). Huma security in theory and practise: application of the human security concept and the United Nations trust Fund for Human security. United Nations trust Fund for Human Development.
8. Interior, M. o. (2014). National Internal security Policy, 2014-2018. Islamabad: Government of Pakistan.
9. Irshad, M. (2011). Terrorism in Pakistan: Causes and Remedies. *The Dialogue*, 225-243.
10. Jahangir, Hamza. 2012. "Security and Human Development in Pakistan." Zurich: E-International Relations.
11. Javaid, U. (2011). Partnership in War on Terror and Mounting militant extremism in Pakistan. South Asian Studies, 227-239.
12. Karim Hussein, Donata Gnisci and Julia Wanjiru. 2004."Security and Human Security: An Overview of Concepts and Initiatives What Implications for West Africa?". Paris: Sahel and West Africa Club.
13. Khan, Ehsan Mehmood. 2013. *Human Security in Pakistan*. Islamabad: Narratives.
14. Khan, Foqia Sadiq. , 2005. "Human Security: A Country Paper on Pakistan." Islamabad: Sustainable Development Policy Institute.
15. Liotta, P. H., & Owen, T. (2006). Why Human security? Whitehead Journal of Diplomacy and International Relations, 37-55.
16. Mary Martin, Taylor Owen. 2010. "The Second Generation of Human Security: Lessons for the UN and EU Experience." *International Affairs* 86(1): 211-24.
17. National Human Development Report. 2009. "Human Security in Pakistan." Islamabad: UNDP.
18. Policy Brief. 2012."Culuturally-Based Peace building in Pakistan." Washington DC: 3P Human Security.
19. Schafer, P. J. (2013). Human and water Security in Israel and Jordan. Environment, Security, Development and Peace, 5-18.
20. Schaffer, H. B., & Schaffer, T. (2011). *How Pakistan negotiates with United States*. Lahore: Vanguard books.
21. Snyder, Craig A. 2012. *Contemporary Security and Strategy*. New York: Palgrave Macmillan
22. Tadjbakhsh, Shahrbanou. , 2005. "Human Security: Concepts and Implications with an Application to Post-Intervention Challenges in Afghanistan." Cambodia: Centre for Peace and Conflict Resolution.

CHAPTER NO. -10

NATIONAL SECURITY; ARMED FORCES OF PAKISTAN

Introduction

"Geography is destiny". It cannot be altered. Together geography and national interest provide reason de attire for the existence of the armed forces of any state. The geostrategic location where offers multidimensional benefit, also pose insurmountable threats. Pakistan inherited geography which has both these characteristics and thus from the very infancy posed wide ranging challenges to the armed forces of the state. Pakistan shares borders with India (2912 Kms), China (523 kms), Afghanistan (2430 kms) and Iran (909 kms), besides 1046 kms long coastal line (CIA World Factbook, 2014). At independence, Pakistan Army, though small undeveloped and understaffed, the force quickly understood the challenges, realised their role and embarked on the path of developing into one of the finest armies of the world. In this regard, the vision and dynamism of the father of the nation Quaid-e-Azam Muhammad Ali Jinnah needs no mention. He truly identified the tasks that lay ahead (base) for the forces and therefore, through his visits, formal and informal interaction provided a vision and guideline to be followed which was grasped by the armed forces of the state in the same letter and spirit. Thus the foundations of strong armed forces as coercive power were laid by the father of the nation at the time of independence. However, there were other factors also that became determinants of the adoption of (towards adopting) such an approach. Being the Armed Forces of the Islamic Republic of Pakistan, the first and foremost guidance was sought and aptly provided by the Quranic Injunction cited above. Second inspiration came from the injustices committed by the boundary commission at the time of partition and the third inspiration came from the unfinished agenda of partition that is Kashmir. Last but not the least the biggest refugee dilemma in the history of the mankind put on the shoulders of the armed forces immense responsibility in the context of both security and management. In the absence of other compatible institutions the armed forces stood out and delivered in the difficult times. In the ensuing period through assistance of China and US, the Armed Forces of Pakistan developed their capability and adopted an institutionalized approach to tackle in hand problems and those that lay ahead. This institutionalized approach earned them the status of the "Centre of Gravity of Pakistan". Today, notwithstanding the shortcomings, role the Armed Forces have played in thwarting all external and internal threats to the integrity and sovereignty of state and ensuring national security bears testimony to the above claim. Having identified the raison de attire, let us now see the importance of the national purpose of a state which if correctly identified becomes the beacon light for entire machinery of the state.

National purpose of any state is that fundamental reason on which the total structure of the state is built. In case of Pakistan's national purpose and the national security thereof have three fundamental components; firstly while retaining the ideological notion, the desire to be part of global community as an accepted member, secondly, ensure sovereignty and territorial integrity against threats of all types and manifestations and thirdly the desire to emerge as a developed state in favorable environment. Besides, other interests could be the desired end sates such as national ethos and values, demos, political stability and economic well-being etc. If above all is considered as true and correct then all the institutions of the state should strategize their aims, objectives and plans honing on to the achievement of above mentioned goals. Divergence from above will certainly lead to chaos, anarchy and finally the security dilemma as identified by Ken Booth and Nicholas J. wheeler in their book The Security Dilemma (Fear Cooperation and Trust in World Politics).

It is in this backdrop that I in the theoretical and historical perspective try to analyze the (in this historical and theoretical perspective, I have tried to analyze) the role of Armed Forces of Pakistan in dealing with the National Security requirements of the state and how this role has transformed post 9/11 scenario and have the armed forces responded to this role correctly or not? In the end, I shall proffer some suggestions to avert any misperceptions and misunderstanding besides providing policy guidelines for a harmonious relationship between the armed forces and other institutions of the state.

Theoretical Foundation

All along its cheered history, Pakistan has been a victim of criticism internally, regionally and globally. The criticism ranged from failure of governance to being labeled as a failed state. In such an environment and under such condition, while discussing the case of Pakistan and her armed forces, I tend to go with the realist and neorealist traditions. The role of military power as an element of national power in such conditions is clear and beyond doubt. Today the international politics revolves around three intellectual perspectives namely realism, liberalism and constructivism.

These perspectives reflect the way the world or a particular state is expected to operate. The policy makers are in effect influenced by the acceptance of the core assumptions of a particular perspective suiting their immediate environment in particular and regional and global environment in general. Amongst these perspectives the oldest and the most dominant is perhaps realism. Intellectual roots of realism date back to the times of Thucydides and Machiavelli. The concept envisages that the world or international system is anarchic and international politics is conflict prone and therefore dangerous in which security cannot be guaranteed. That the states are primary unitary and rational actors and their ultimate objective is survival and attainment of maximum power. Thus the only way to ensure sustainable security is by adopting self help strategies. In doing so the states may according to the prevalent environment decide to either achieve them through internal or external or combination of both the means. In case of the first one the states will internally strengthen own military, political and economic potential etc, while in case of the later it may look forward to alliances and international organizations. And in the last case the states may adopt both the strategies concurrently and endeavor to achieve sustainable peace and development.

Realist and neorealist contend that state is the main actor and that international system is anarchic (Morgenthau & Nations, 1948). Thus in an environment of international anarchy, states are forced to rely on their own resources in order to ensure their continued existence. Military power is thus the ultimo ratio of state power. Armed forces exist to defend the state against real or potential external threats and as a coercive tool to promote and protect national interests abroad and if required even at home.

So then what the national security is? This is one question that if answered properly paves way for further discourse. The term national security over decades has transformed from the most conventional nature into including security against numerous traditional and non-traditional threats. Today therefore as explained in the earlier chapters also it encompasses ensuring territorial integrity of the state from external as well as internal threats, safeguarding the people, their rights and interests (individual freedom) and their way of life. Today it also includes economic prosperity and political stability. Therefore, preserving national security of the Islamic Republic of Pakistan would require safeguarding the borders, individual and collective freedom including other national and religious values, the constitution, law of the land and institutions established to protect them. In essence then any controversy or difference of opinion on national interest cannot undermine the basic purpose of any state and that is the protection of core interests and continued vitality as a cohesive entity amongst the comity of nations.

Having settled with the above definition, then what is the challenge to this concept of national security. Today in my humble opinion the biggest challenge posed to the traditional concept is the over arching influence of the nontraditional threats to national security such as political security, economic security, energy security, human security, narcotics, smuggling, and environmental security etc. Notwithstanding the realisation that the abovementioned threats do impinge upon the overall security canvas of a state yet these issues deserve a place down the ladder below traditional security agenda. To be more clear the survival of state should always be the single most important objective of any state. If any state loses such a capability it does so to its own peril.

The security challenge, not the security dilemma faced by Pakistan is multi faceted and multi dimensional. Why I call it security challenge is that element of doubt is nonexistent and the intentions of the adversary are clear and beyond doubt. The recent joint declaration by the leaders of the state that RAW (Research and Analysis Wing) India's prime intelligence organization is deeply involved in destabilizing Pakistan by employing all possible means is enough to explain the gravity of security challenge. Besides the recent episode of attacks along the working boundary in Sialkot sector and along the line of control is true manifestation of this security challenge. Also, ever since independence, Pakistan has continuously faced existential threat. India has never accepted Pakistan as an independent state. Thus it has been India's continuous endeavour to not only involve Pakistan in various disputes and problems but also portray Pakistan as a rogue, militarised and failed state. The underlying purpose of such a strategy has been to weaken Pakistan internally and isolate it globally and then either makes it a satellite state or a submissive neighbor; thereby achieving her permanent regional and global aspirations world over.

Samuel P Huntington in his book "The soldier and the State" writes that the military institutions of any society are shaped by two forces: a **functional imperative** stemming from the threats to the society's security and a **societal imperative** arising from the social forces, ideologies, and institutions dominant within the society. Military institutions which reflect only social values may be incapable of performing effectively their military functions. On the other hand, it may be impossible to contain within society military institutions shaped only by functional imperatives (p.2.(Huntington, 1957). Thus for the militaries to be more effective and efficient **both these imperatives have to be taken into consideration**. We will talk more about these imperatives with specific reference to Pakistan in the section defining threat matrix.

Another important aspect that I must clarify at the outset is that during the process of research, I scanned many books and articles written by eminent scholars and researchers on Pakistan and the role of Pakistan Armed Forces. The agenda and the focus of majority of the literature on the subject has been trying to find the short comings of the armed forces and blaming them for all the ills today prevalent in the society. A few to mention are; Pakistan Army is responsible for disallowing the constitutional supremacy, it is Pakistan Army that has prevented development of the state institutions, Pakistan Army has been described as an Enterprise always trying to safeguard her interests, Pakistan Army has lust for power and any effort to democratise the role of Armed Forces has invited martial law and Pakistan Army is responsible for the dismemberment of the country in 1971etc. My contention in the chapter is totally different. I am not into blame throwing game. I just want to highlight that the state of Pakistan or for that matter any other state is like an engine; malfunctioning of any or one part results into mechanical failure. The gravity of such a failure is determined by the role and importance that part. It is the responsibility of all the state organs and it is the unity of effort of all these organs that can bring sustainable prosperity and development in any state of the world. Thus it is the mutual failure of all or most of the important state organs to bring about positive change in the country so as to harness the true potential of the state.

I hereafter will approach the subject in the same context. Now the questions that strike any intriguing mind are that has the role of the Armed Forces of Pakistan been in line with the above mentioned concepts? Secondly what alternate is available to enhance the overall security potential of the country? And thirdly what should then be the course of action on part of the democratic government to develop a well articulated national security policy for a prosperous future?

Coming to the first question that is whether or not the role of the armed forces has been in line with the above mentioned concept. My short answer is both yes and no. However, when viewed in broader plane "yes". The armed forces of Pakistan have been identified as the centre of gravity of the state power. Weakening the COG can weaken the state. In the same context an overarching argument against my contention will be dismemberment of Pakistan in 1971 and Samuel P Huntington in his book "The Soldier and the State" aptly describes the aim of the national security policy and role of the armed forces. He contends:-

The aim of the national security policy is to enhance the safety of the nation's social, economic and the political institutions against threats arising from the other independent state. ***National security policy may be thought of as existing in three forms and on two levels.Military security policy*** *is the program of activities designed to minimise or neutralise efforts to weaken or destroy the nation by armed forces operating from outside its institutional and territorial concerns.* ***Internal security policy*** *deals with the threats of subversion- the effort to weaken or destroy the state by forces operating within its territorial and institutional confines.* ***Situational security policy*** *is concerned with the threat of erosion resulting from long term changes in social, economic demographic and political conditions tending to reduce the relative power of the state. Each of these three forms of policy has an* ***operating level and an institutional level. Operating policy*** *consists of the immediate means taken to meet the security threat.* ***Institutional policy*** *deals with the manner in which operational policy is formulated and executed. Civil-military relations is the principal* ***institutional component of military security policy.*** *The immediate operating issues of military policy normally involve (Huntington, 1957)*

- *The qualitative issues of the size, recruitment, and supply of the military forces, including the fundamental question of the proportion of state resources devoted to military needs*
- *The qualitative issues of the organisation, composition, equipment, and deployment of military forces, including the types of armament and weapons, the location of bases, arrangements with allies and similar questions; and*
- *The dynamic issues of the utilisation of military forces: when and under what circumstances force is brought into action.*

However, even before we try to answer the two questions let us see how the threat matrix has transformed over the decades following independence. As only the correct assessment of the threat alone can lead us to right conclusions regarding development and implementation of national security strategy.

Transforming Threat Matrix

Armed forces are a near-ubiquitous feature of the state of Pakistan as is the threat. Ever since independence, existential threat has been omni-present. The paradox of the creation of Pakistan has been the fact that we all along our independent history have never faced a security dilemma. Rather our concern has been the all alive and real security challenge posed by our immediate and distant neighbors implying elimination of the degree of uncertainty. Armed Forces of Pakistan therefore, remain key actors in addressing the national, regional and global insecurities. In this section we will relate the threat matrix with the two imperatives highlighted in the earlier pages and provided by Samuel P Huntington. The first imperative i.e. functional imperative, I argue that it is an undisputed reality that the society's security is threatened by both the east and the west. From the east by the hegemonistic (hegemonic) designs of India and from the west by the fragility of the state of Afghanistan that all along it chequered history has been invaded by foreign powers resulting into tribal nature of political system and unacceptability of any central government save a brief period. Besides, erstwhile Soviet Union also posed serious existential threat to Pakistan when it invaded Afghanistan probably with an intention to have access to the warm waters of the Indian Ocean through Arabian Sea in Pakistan. However, the threat subsided with defeat of Soviet Union in Afghanistan and consequently her disintegration.

During the earlier years the existential threat was posed by India alone, post 9/11 the threat expanded along both horizontal and vertical planes. The dynamic nature of the threat and insecurities therefore necessitated a fundamental reassessment of the role, capacity and capabilities of armed forces vis-à-vis their earlier traditional composition and capabilities. In this changed dimension another question that came up was that whether or not they are the most appropriate organization to do so? As part of the transformation process armed forces of Pakistan increasingly identified and embraced on new roles in order to not only secure the state but also to legitimize their institutional existence. In some ways these roles were already loosely attached to the armed forces especially the land forces, however, due to the overarching conventional external threat received little attention by the military leadership and thus remained dormant. Although as early as 1980s, ripples from societal imperative started effecting the law and order of the country, however, it was considered to be well within controllable limits of the police, second line forces duly supplemented by the first line forces and the rudimentary technological provisioning of the leftist and outlaws.

Post 9/11 the threat matrix transformed profoundly. The societal imperative came into play with an unprecedented intensity. This impacted upon the armed forces in three distinct manners. Firstly, the role expanded horizontally to include the western border, secondly, the war on terror and the presence of extra regional forces close to our borders and in international waters in the Indian Ocean and Arabian Sea directly and indirectly impinged upon the security of the state and thirdly, the internal law and order situation arising from the failure of state polices and foreign direct and indirect involvement taxed the at hand sources of the armed forces; posing serious challenge to their capabilities then mostly of traditional nature. Lastly, the use of technologically advanced equipment outclassed the conventional equipment held with the armed Forces and thus posed a vertical challenge.

These three conditions led to the policy makers and professionals to conclude that besides the conventional role, military responses entailed renewed emphasis on the counterinsurgency and internal security functions of the armed forces. Secondly, the demands of the war on terror challenged the position of the regular armed forces as the main organization for meeting external security threats to the state. Thirdly, besides capacity building of the regular armed forces, second line and paramilitary forces also require capacity building and modernization.

Notwithstanding above, the Armed Forces had to navigate an uphill and difficult course. That is to say that it must strike a fine balance between both the probability and imminence of the threats it faces and cost incurred to neutralize the threats. The threat matrix as highlighted earlier now has expanded to include nontraditional threats also, thus actually the threats instead of shrinking have increased manifold and grown across both horizontal and vertical planes. Thus while the actions should be taken the investments require to be strategic and selective. Such actions should also prevent public sensitivities from heightening about their role in internal law enforcement and in stability of the fragile democratic structure. We witnessed that both the Government and military were quick to realize these challenges. Thus together despite some economically difficult times embarked upon raising a couple of new Frontier Corps, increasing the strength and capability of police, modernizing the rangers and re orienteering their training and equipment needs to suit the requirements of the compounded threat.

The above concept to me does not seem to be new. If analyzed we realize that it actually existed in a skeleton and crude form during the infancy of the state. The manifestation of this concept lay in the civil defence, national guards, basic military training imparted at college level, razakars and levy in yester NWFP and Baluchistan. However, all these had to be abandoned partially due to lack of funds and partially due to non-development of a wholesome and mature security concept.

Thus it may not be wrong to conclude that although to greater extent military security policy was modified and requisite doctrinal changes incorporated, however, at the state level lack of seriousness displayed by absence of nationalist approach left the internal and institutional security policies in disarray. Instead the efforts were put in to promote own political interests.

Current National Security Apparatus

Before I take on the existent security apparatus of the country for building on my argument on the security structure of the state, I take support of the constitution of Pakistan which is the supreme and the most sacred document abrogating which tantamount to treason. For the purpose of clarity, I searched for the national purpose of the state as I found in the constitution of the United States of America has been highlighted in earlier chapter. The national purpose of Pakistan as a clear separate heading is non-existent, however, it to some extent does exist in the form of preamble and linking it to the Objectives Resolution. The main points of both the preamble of the Constitution of the Islamic republic of Pakistan and the Objectives Resolution are summarized below:

- Sovereignty over the entire universe belongs to Almighty Allah alone, and the authority to be exercised by the people of Pakistan within the limits prescribed by Him is a sacred trust.

- The state shall exercise its power and authority through the chosen representatives of the people.
- The principles of democracy, freedom, equality, tolerance and social justice as enunciated by Islam shall be fully observed
- Muslim shall be enabled to order their lives in the individual and collective spheres in accordance with the teachings of Islam as set out in the Quran and Sunnah.
- Adequate provision shall be made for the minorities to freely profess and practice their religions and develop their cultures
- Pakistan shall be a federation wherein the units shall be autonomous with such boundaries and limitations on their powers and authority as may be prescribed.
- Fundamental rights shall be guaranteed.
- The judiciary shall be independent.
- Integrity of the territories of the Federation, its independence and all its rights, including its sovereign rights on land, sea and air shall be safeguarded.
- In order for the people of Pakistan to prosper and attain their rightful and honored place amongst the nations of the World and make their contribution towards international peace and progress and happiness of humanity.
- Pakistan will be a democratic state based on the principles of social justice.

Later Articles 243-245 of the constitution of the Islamic Republic of Pakistan delineate the command structure and role of the Armed Forces. According to the constitution "The Federal Government shall have control and command of the Armed Forces". It further says that without any prejudice to the generality of the foregoing provision, the Supreme Command of the Armed Forces shall vest in the President. The President shall subject to law, have power to firstly raise and maintain the Military, naval and Air Forces of Pakistan; and the reserves of such forces; and secondly, to grant commission in such forces. The President shall on advice of the Prime Minister, appoint the Chairman, Joint Chiefs of Staff Committee, the Chief of the Army Staff, the Chief of the naval Staff; and the Chief of the Naval Staff, and also shall determine their salaries and allowances. In article 245 of the constitution the role of the armed forces is given as following;

The Armed Forces shall under the directions of the Federal Government defend Pakistan against external aggression or threat to war, and, subject to law, act in aid of civil power when called upon to do so. The validity of any direction issued by the Federal Government under the clause (1) shall not be called in question in any court. A High Court shall not exercise any jurisdiction under Article 199 in relation to any area in which the Armed Forces of Pakistan are, for the time being, acting in aid of civil power in pursuance of Article 245. Provided that this clause shall not be deemed to affect the jurisdiction of the High Court in respect of any proceeding pending immediately before the day on which the Armed Forces start acting in aid of civil power. Any proceeding in relation to an area referred to in clause (3) instituted on or after the day the Armed Forces start acting in aid of civil power and pending in any High Court shall remain suspended for the period during which the Armed Forces are so acting.

Dispassionate analysis of the preamble and the articles 243-245 lead us to the conclusion that:-

- The national purpose is vague and if not vague then not very clearly articulated to derive from it the national objectives and consequent national strategies. The national purpose should make it binding on the state institutions to define for them the national strategies, mission and goals according to the dictates of the changing environment.

- The second point I want to make is that although in accordance with the parliamentary form of the Government head of the Armed Forces is the President of Pakistan who also has the responsibility to raise and maintain the armed forces, however, the operational control rests with the federal government and by virtue of that with the Prime Minster. The above contradiction leads to weaknesses such as:-

    - President is just a figurative or dummy head of the most important state organ without any real powers. Prime Minister is bound to keep the President informed only. Balance of powe' thus tilts in favour of the Prime Minister, making President only a ceremonial figure.

    - President does not have the powers to dictate any measures for meeting the requirements of the armed forces.

- Prime Minister has to seek consent and approval of the President to appoint the services chiefs. Also without such a notification the orders of the Prime Minister bear no meanings. The case in point is the dismissal of the ex COAS General Pervez Musharaff and appointment of the Lieutenant General Zia ud Din Butt (Retired) as COAS. In such an eventuality actions of Prime Minister become un parliamentary and illegal.

- According to the Article 232 of the Constitution of Islamic Republic of Pakistan, President and not the Prime Minister can proclaim emergency. Such ambiguities and contradictions although may not be of any significance during the periods of peace and routine functioning, however, in emergencies these do surface and impact negatively on the overall decision making process thus resulting into either delayed or else faulty decisions and ultimately the burden of responsibility.

Having highlighted my observation about the constitutional anomaly in governing the armed forces, let us now have a look on as to how the state security network is structured and as how it is being governed. Broadly speaking the security organizations in Pakistan are divided into two major components firstly the ones that are governed by and operate under the Ministry of Interior and the others that are governed by and operate under the Ministry of Defense. The organogram of both organizations are as following:-

Ministry of Defense

Ministry of Interior

Although the two organizations are comprehensive and contain all the elements required to become an effective and deterrent organizations in being yet what they lack is coordination, cohesiveness and unity of command essentially required during the times of emergencies and a central and efficient intelligence system.

Notwithstanding above during the term of current government, a ray of hope has been kindled both by political as well as military leadership of the state. Never before in the history of Pakistan, the two leaderships tangents apart have been seen operating so close and in such a coordinated manner expressing common resolve to eliminate the menace of terrorism, instability, sectarianism, militancy and lawlessness. Political leadership has to have confidence in the military leadership so as to ensure a strong state. Approval of NAP (National Action Plan), launching of the operations clean up in Karachi and FATA and coordinated joint conferences and statements are some of the positive steps that are being taken to ensure national harmony. Lieutenant General Naveed Mukhtar, Commander 5 Corps address on 16 May 15, however, made it crystal clear that the ongoing operation is being hampered by the political expediencies and vested interests of a few. Thus underlying conclusion of such remarks is sufficient evidence to conclude that much more needs to be done to achieve long lasting sustainable results. It is in this backdrop that I feel following steps if taken shall go a long way in ensuring sustainable peace and development.

Findings and Recommendations

- **Geography and Neighbours a Universal Reality.** As contented earlier geography is constant and cannot be altered and so are the neighbors. Both are therefore a geo political and geostrategic reality. It therefore, becomes incumbent upon us to remain alive about the hegemonistic (hegemonic) designs of our eastern neighbor and others on our western borders who defy the international border. The notion that the threat from the eastern borders has considerably reduced is a farce. The latest statements of Indian Premier Narindara Modi and continuous violation of the cease fire across the working boundary should be sufficient to open the eyes of policy makers in Pakistan. The strategic thinking of the Indian strategists even today has foundations in the writings of the Chankiya Kutilya advisor to Chandra Gupta Maurya. In his volume "Arthashastra" he advises the policy makers that firstly the state should have an unflinching resolve to increase power and territory and secondly while remaining cautious to the intentions and capabilities of the neighbors success lies in befriending the neighbors' neighbor. The practical manifestation of both these have been repeatedly witnessed by Pakistan all along the chequered history of our relationship especially in case of India's approach in settling the festering disputes of Kashmir, Siachen, Sir Creek and water distribution; and her growing relationship with Afghanistan and Iran in particular and Central Asian Republics in general. Given these circumstances Pakistan will have to understand that the border and geography cannot be altered and therefore in order to avert the hegemonic designs of India she has to adopt an aggressive, firm and sustainable policy both at home and internationally to pressurize the Indian Government to address the long standing issues. In this regard it must be remembered that in order to achieve this goal the state has to build on and strengthen all the elements of national power in a prudent manner.

- **Regionalism and Strong Regional Alliance a Necessity Rather Than a Luxury.**

- **Downsizing Armed Forces not a Viable Option.** During the last decade US has been consistently urging Pakistan to reduce the size of the armed forces due to the nuclear deterrence and the result balance of threat. However, as we have already analyzed in the earlier and this chapter that the armed forces their existence and strength is always contingent upon the threat, capability and intentions of the adversaries. We have concluded that the functional as well as societal imperatives dictate that instead of downsizing we should not only enhance the qualitative capabilities of the conventional armed forces meant to neutralise and thwart any existential threat but also strengthen the second line forces on Homeland security concept being in vogue in the world. More will be covered on this in the next chapter. Moreover, in addition to strengthening and enhancing the capabilities of the armed forces across both horizontal and vertical planes, economic prosperity needs to be commensurately built up. Dr Muhammad Yaqub has rightly identified that "No country has been able to maintain and sustain a deterrent defence and an independent foreign policy with a weak economy that becomes increasingly dependent on foreign handouts".

- **Adoption of the Concept of Homeland Security.** One of the major weaknesses identified during the course of research has beenabsence of effective checks and balances in the society which led to the sorry state of law and order and nurturing of the notion of getting away with any wrongdoing. Most of us have been in west and travelled vastly. Do we want to conclude that the humans living in those parts of the world are angles and possess superior IQ. No their potentials have harnessed through development of necessary systems. In my opinion in Pakistan also this notion has been able to take roots because of two reasons one failure to develop proper internal law and order maintaining structure and secondly due to the failure of first speedy dispensation of justice. After all humans by nature are prone to defiance and always look for escapist approach. USA, UK and Europe having experienced the gravity of the ever growing threat and resultant challenges adopted the subject concept. The concept in a crude form does exist in Pakistan however, in order to be effective it has to refined, developed and the capabilities of the existing and new organizations structured built to suit the dilates of the time and ever transforming challenges.

- **Democratic Civilian Control of the Armed Forces.** Democratic civilian control of the armed forces is not only preventing the armed forces from seizing power but essentially it is aligning the political and military goals. That is to say that the interests of the military and the political leadership both do not overtake the broader national and societal interests. Going a little further it means that military should not subvert the democratic constitutional authority or to absorb a disproportionate amount of resources relative to other societal values and priorities, concurrently ensuring that the needs of the military to fulfill the task assigned are adequately met in terms of both men and material. Thus it is actually the responsibility of the political leadership of the country to decide the national goals and objectives and ensuing national security strategy. However, it is of utmost importance to ensure that the goals are compatible to the capability and the needs are met according to the desired capability of the armed forces. In doing so then the concurrence of the military leadership and their priorities towards attainment of the laid down national objectives should be given due diligence and importance.

  The above concept in no way suggests undermining the strength, power or required professional independence of the military as an institution. As unity of command and obedience to the military leaders is the essence of military training and direly needed in the wake of emergency and war. It is very much possible to strike a fine balance between the two by drawing a clear line dividing have and have not. True manifestation of the concept is visible in the NATO and Europe where it has become a prerequisite for the states aspiring for the membership of NATO. The concept of political non intervention by the armed forces is engrained in the democratic theory that is that no one individual or a group represents the will of the entire population or in other words national will, and that only the democratically elected governments can speak for the nation as a whole. Thus on part of the military and related professionalism it implies political subordination to the policy decisions and specified objectives set forth by the constitutionally elected representative.

In this regard Lutwak rightly identifies seven areas as the domain of the constitutionally elected government which include firstly, defining the security threats, secondly determining when the force will be used, thirdly strategic planning, fourthly allocation of resources to the defense (defense budget), fifthly defining the mission of the armed forces, sixthly deciding on the force structure development and procurement and lastly senior military promotions. He contends that anything short of civilian supremacy is an incomplete democracy (Luttwak, 1999, p. 99).

There are dangers to the democratic civilian military control of the armed forces too. Such dangers and their negative implications are visible even in advanced European democracies. At time as part of political expediency short term quick fix solutions are imposed on the armed forces which become detrimental to the force structure, cohesion and adversely affect the force structure development. It takes decades to develop the capability of the armed forces that can be destroyed as a result of one wrong policy decision to gain political benefits. Therefore, it is the responsibility of the civilian government that while ensuring accountability, give full leverage to the armed forces for their professional development and mission oriented training. Thus the concept not only envisages civilian supremacy, but also the courage and veracity to tackle the military issues seriously with utmost professionalism and astuteness.

On the other end of the pendulum military also shares the burden of responsibility. Military should provide professional expertise and advice to the civilian leaders to reach at the correct decisions in an expedient manner without getting politicized which is also the responsibility of the civilian government to prevent any such notion from developing.

To conclude the argument it does not mean that differences will not erupt in the process of decision making. Even in most advance democracies where the system are well established and have matured over decades, differences did appear as was witnessed during Vietnam War (civil-military gap in US), Somalia affair in Canada, ISAF in Afghanistan and numerous others. However, notwithstanding above success of the concept lies in realistic division of power, development of structure and mechanism to avert any one party from asserting undue influence, honesty of purpose, national spirit and above all professionalism, understanding and commitment on part of both the sides.

Conclusion

Pakistan today faces assortment of functional and societal challenges which with every passing day are developing, increasing and transforming. The direct threats range from conventional limited war based on Indian PAO strategy to use of nuclear, biological and chemical weapons besides creating a two front dilemma. Isolation of Pakistan and imposing sanctions in case of defiance and or resistance to the global power aspirations are not a far reached conclusion. In the realm of societal imperative Pakistan in the near and intermediate future is likely to continue remains unstable due to to the ethnic and sectarian divide and fragile political structure. The economic and developmental challenges are also likely to continue haunting the state unless a well considered policy is not only chalked out but also immediately executed on war footings. The law and order state is most likely to remain poor in the foreseeable future. The strong institutional level which is vital for the success of operational level is also likely to remain weak due to political instability and lack of realization on part of the policy makers. In such a scenario in order to protect the country against internal and external existential threat existence of potent, professionally competent and technologically advanced and strong armed forces are vital. However, in order to be successful the armed forces of Pakistan require a clear direction and autonomy. The direction has to be chalked out by the democratic government in the form of a comprehensive and all allusive National Purpose and National Security Policy. The policy if properly available will sufficiently harness the armed forces and act as a beacon light for all their current and future actions. Also the loose ends available in the constitution should be reanalyzed and modified to prevent any future controversy. Grant of autonomy to the armed forces or for that matter all other state institutions of the same stature may appear to be contentious. However, the notion of professional autonomy has merit. Whereas armed forces should be made accountable and transparent, provision of professional autonomy will make the institution of armed forces more capable and inspire more loyalty if there is room for the military leadership in development and application of professional knowledge gained over decades and in the exercise of discretionary judgment.

Lastly in the wake of existing political turmoil I leave a food for thought for the nation with Samuel P Huntington idea that he floated for the American society in the concluding chapter of his book "The soldier and the state" that in the wake of omni- present functional and societal challenges, I also propose that the Pakistani society should become less liberal and more like military in its culture and values.

References

1. Huntington, S. P. (1957). *The soldier and the state: The theory and politics of civil-military relations* (Vol. 514): Harvard University Press.
2. Morgenthau, H., & Nations, P. A. (1948). The struggle for power and peace. *Nova York, Alfred Kopf.*

ANNEXURES

Operation Zarb-e-Azb

Introduction

Operation Zarb-e-Azb is a joint military offensive being conducted by the Pakistan Armed Forces against various militant groups, including the Tehrik-i-Taliban Pakistan (TTP), the Islamic Movement of Uzbekistan, the East Turkestan Islamic Movement, Lashkar-e-Jhangvi, al-Qaeda, Jundallah and the Haqqani network. The operation was launched on 15 June 2014 in North Waziristan along the Pakistan-Afghanistan border as a renewed effort against militancy in the wake of the 8 June attack on Jinnah International Airport in Karachi, for which the TTP and the IMU claimed responsibility. Part of the ongoing war in North-West Pakistan, up to 30,000 Pakistani soldiers are involved in Zarb-e-Azb, described as a "comprehensive operation" to flush out all foreign and local militants hiding in North Waziristan. The operation has received widespread support from the Pakistani political, defense and civilian sectors. As a consequence, overall security situation improved and terrorist attacks in Pakistan dropped to a six-year low since 2008.*Zarb-e-Azb* means "sharp and cutting strike". *Azb* also refers to the sword owned by Hazrat Muhammad (PBUH), which he used in the battles of Badr and Uhud.[1]

Backdrop

Peace negotiations with the Taliban were announced by Pakistani Prime Minister Nawaz Sharif after his election, although previous attempts to engage TTP in dialogue had failed. The first session of talks, between committees appointed by the Pakistani Government and the Taliban, was held on 26 March 2014 at Khyber Pakhtunkhwa House in Islamabad. The Taliban did not name representatives from their ranks, instead nominating pro-Taliban religious figures to present their views. The terrorists called for the implementation of Sharia in Pakistan; the Government of Pakistan demanded the cessation of hostilities, insisting that talks be held within the framework of the Pakistani constitution. A month-long ceasefire was reached on 1 March 2014. Besides the meetings at Khyber Pakhtunkhwa House, negotiations also involved helicopter travel by government representatives to the areas under militant control near the Pakistan-Afghanistan border. The government had indicated that stronger military action would be implemented if the talks failed (Saleem, 2014).

Failure

Negotiations collapsed after the execution of 23 Pakistani Frontier Corps soldiers by the Taliban on 17 February 2014. The soldiers had been held by the insurgents since 2010, and on 17 April 2014 the TTP formally ended the ceasefire. Taliban infighting since March 2014 killed more than 90 militants. The strife, triggered by differences between the Mehsud group (led by Sheheryar Mehsud) and another TTP faction (led by Khan Said Sajna), impeded the negotiations.[46] The negotiations were irreversibly damaged by a terrorist attack on Karachi Airport for which the Taliban claimed responsibility and which killed 28 people (including security personnel). A Pakistani military official was quoted to have said, "The army is ready for an operation. It now all depends on the government to make a decision" (Khan, 2014).

Jinnah Airport attack

The operation began one week after a terrorist attack on Pakistan's busiest airport. On 8 June 2014, 10 militants from the Islamic Movement of Uzbekistan and TTP attackedJinnah International Airport, Karachi, killing 28 people including security personnel and wounding at least 18. In retaliation, the Pakistani military launched a series of air-strikes targeting terrorist hideouts in the areas bordering Afghanistan. At least 25 militants were killed on 10 June. The figure also included foreign militants killed (The News, 2014). Two drone attacks on 12 June killed Uzbek, Afghan and local militants. On 15 June, the Pakistani military intensified air-strikes and bombed eight foreign militant hideouts, killing as many as 140 militants (most Uzbek, including persons linked to the airport attack and airport attack commander and mastermind Abu Abdur Rehman Almani) in North Waziristan (2014).

Timeline

(a). 2014

15 June 2014

The first phase of the operation began with intensified airstrikes in North Waziristan, targeting militant training facilities, hideouts, and other infrastructure (The News International, 2014). The Pakistani military destroyed eight militant hideouts in the previous night's airstrikes. Hideouts in Degan-Boya and Datta Khel were targeted by jet aircraft, since foreign and local insurgents linked to the Karachi airport attack were confirmed there; an ammunition dump was also destroyed. As many as 140 militants (mostly Uzbek) were reportedly killed in the strikes, including commander and airport attack mastermind Abu Abdur Rehman Almani. North Waziristan was sealed by troops on its border with neighbouring agencies and FATA regions to block the movement of militants. In North Waziristan, troops cordoned off insurgent bases (including those in the towns of Mir Ali and Miranshah).[55] Logistical and administrative arrangements for IDPs were made by the Disaster Management Agency, with the establishment of registration points and IDP camps. Surrender points were established for militants wishing to give up their arms (Times of India, 2014). Aerial surveillance of the area was conducted. Afghan security forces were requested to seal the border on their side to prevent militants from escaping across the border and initiate immediate measures to eliminate TTP militants and their hideouts in Kunar, Nuristan and elsewhere in Afghanistan (2014).

16 June 2014

Seven fleeing militants were killed on the outskirts of Mirali overnight, with three soldiers injured in the exchange of fire. In a separate incident, seven more militants were killed when they tried to flee from the cordoned-off area. Two Pakistani soldiers were reported killed in an exchange of gunfire. Six militant hideouts in Shawal, North Waziristan were destroyed by an early-morning airstrike by two fighter jets, with 27 militants killed. There were no civilians in Shawal Inter-Services Public Relations(ISPR) reported six soldiers killed and three injured in an improvised explosive device (IED) explosion between the Afghan border and Ghulam Khan Tehsil in North Waziristan. According to ISPR, a convoy of security forces was targeted on Bane Dar Road in Ghulam Khan (on the Pakistani-Afghan border). Forces cordoned off the area, launching a search operation. Three insurgents were killed by Special

Services Group sniper fire while planting IEDs near Miramshah (IMU, 2014).

17 June 2014

Airstrikes destroyed six hideouts in North Waziristan, killing 25 foreign and local militants. Airstrikes were also conducted in the Hasokhel area of Mir Ali. At least three suspected militants were killed when they attempted to flee a cordoned-off area in Miramshah, with one soldier injured in the exchange of fire. No operations had begun in developed areas to ensure that no militants could escape the cordon and all verified civilians were evacuated. More than 40 percent of North Waziristan was cleared of militants in the first three days of the operation (Walsh, 2014).

19 June 2014

Fifteen militants were reportedly killed overnight in the Zartatangi Heights, east of Miramshah, by army Cobra helicopter gunships. The area was one of the main communications centres of the insurgents. In a separate incident, eight Uzbek militants were killed near Miramshah while planting IEDs on the road between Miramshah and Mir Ali. The evacuation of the civilian population from Miramshah and Ghulam Khan began. Checkpoints were established in a number of locations, where IDPs received food and medicine from security forces.[67] As many as 400 Afghan families left North Waziristan for Afghanistan via Ghulam Khan (Shah, 2014).

20 June 2014

Three militant hideouts in the Qutab Khel area, on the outskirts of Miramshah, were destroyed in the early morning by army Cobra helicopter gunships with artillery and sniper assistance. Twelve militants, including some foreigners, were killed in the strikes and a large cache of arms and ammunition destroyed. Militants inside cordoned-off areas attempted to flee. Six attempts were halted overnight, and three local residents without identification were arrested trying to flee from the cordon. Another 24 suspects disguised as IDPs were arrested at security checkpoints in Mirali and Miramshah. The civilian evacuation continued from North Waziristan towards Bannu, with 200,000 IDPs evacuated so far (Pakistan Tribe, 2014).

21 June 2014

A total of 30 militants were killed in early-morning airstrikes in uninhabited areas of Khyber and North Waziristan Agencies. Jet aircraft destroyed two hideouts near the Pakistan-Afghanistan border in Khyber

Agency, killing 10 militants. These surgical strikes were in line with the security forces' strategy to take on the militants across the FATA (2014). Three hideouts were destroyed in Hassu Khel, North Waziristan, killing 20 militants.

23 June 2014

Eight militant hideouts near Mir Ali were destroyed by jet aircraft during the early morning, killing 15. Tunnels were spotted in the targeted areas. Ten militants were killed attempting to flee from cordoned-off bases in Spinwam and Mir Ali, with two Pakistani soldiers also reportedly killed in an exchange of fire. The curfew was lifted for two hours to evacuate the remaining civilians. At the Saidgai security checkpoint, 414,429 IDPs were registered to date. An army medical corps field hospital was being established in Bannu for displaced persons The civilian evacuation was almost complete. About 450,000 IDPs arrived in Bannu, and were registered at the Saidgai checkpoint (ISPR, 2014).

24 June 2014

Twenty militants were killed and 12 hideouts destroyed in early-morning air strikes in Khyber Agency. Twenty-seven militants were killed in afternoon jet strikes in Mir Ali and the surrounding area, with 11 hideouts and a large weapons cache destroyed. An afternoon suicide car-bomb attempt was foiled in the Spinwam area of North Waziristan. When an explosive-laden vehicle approached a checkpoint outside a civilian hospital, soldiers fired at the vehicle (which exploded 100 meters from the checkpoint. Two soldiers and one civilian were reportedly killed when the roof of a nearby building collapsed from the explosion (Gul, 2014).

25 June 2014

Pakistan Air Force (PAF) jets destroyed five hideouts in Mir Ali, killing 13 militants. Twelve militants surrendered to Pakistani forces (Saleem, 2014).

26 June 2014

The evacuation of 450,000 civilians was completed, and the operation's second phase began with a ground offensive by the Pakistani military. Seven militants surrendered at the North Waziristan surrender center, bringing the total to 19 (Dawn, 2014).

27 June 2014

PAF aircraft destroyed six confirmed hideouts on the outskirts of Mir Ali in an evening raid, killing 11 militants (The Express Tribune, 2014). TTP

Miranshah Commander Umer was killed on the outskirts of the town by security forces that night.

28 June 2014

Militant groups were targeted during the early morning by integrated Pakistani artillery, tank and heavy-weapons fire outside Miranshah, killing seven. An al-Qaeda commander, revealed by initial interrogation as an explosives, IED and suicide-belt expert, was arrested trying to flee from a surrounded base.[76] Three militants were arrested by security forces while trying to cross the Indus River near Mianwali. All river crossings were fortified to seal escape routes (Dawn, 2014).

29 June 2014

Sixteen militants were killed when PAF fighter jets targeted their hideouts in Mir Ali, North Waziristan. According to military sources, seven militant hideouts, explosives and ammunition dumps were also destroyed in the airstrikes (ISPR, 2014).

30 June 2014

Early-morning ground operations began in and around Miramshah. Search operations were conducted by infantry and the Special Services Group, killing 15 militants. Troops discovered underground tunnels and IED-preparation factories in the cleared areas, with three soldiers reportedly injured in an exchange of fire. The civilian population had been evacuated.[80] Since the beginning of the operation 376 militants were killed and 19 surrendered, with 61 hideouts destroyed in the operation's first phase. Seventeen soldiers were reportedly killed (Dawn, 2014).

July

1 July 2014

According to the ISPR, during ground operations in Miranshah a landmine factory was discovered and 225 cylinders, 700 pipes filled with explosive materials and 150 unfinished land mines were recovered. Two Pakistani soldiers were killed and a third reportedly injured when militants ambushed a military vehicle in Mir Ali, North Waziristan (Dunya News, 2014)..

2 July 2014

Ten militants were killed when Pakistani helicopters shelled hideouts in the Khar Warsak region, 12 kilometres (7.5 mi) north of Miramshah, with three hideouts destroyed in the airstrikes (Khan, 2014).

3 July 2014
The bodies of seven militants were recovered in the Darpa Khel area of Mir Ali. A security official reportedly said, "Bodies were of suspected militants who were gunned down by security forces in Mir Ali tehsil of the agency", adding that the forces had advanced from Mir Ali and Miramshah Bazaar towards the outskirts of the agency after destroying three hideouts in Mir Ali (one of which held foreign militants) (Gul, 2014).

4 July 2014
Pakistani soldier Niak Fiaz Mohammad was killed by an IED explosion, during a house-to-house search operation in North Waziristan. Mohammed's funeral was held in Bannu before the soldier's body was transported to his hometown, Mansehra. ISPR director-general Major General Asim Saleem Bajwa said, "We salute the courage and bravery of Niak Fiaz Shaheed" (IMU,2014). A Pakistani soldier from Azad Kashmir was killed in the Jord region of North Waziristan. The soldier, Sadheer Ismail from Malsi Ghari Dupatta, was buried in his family cemetery. Pakistan Army personnel attended Ismail's funeral, where friends praised his determination and resolve.

5 July 2014
Airstrikes targeted Miramshah and the village of Boya, destroying five militant hideouts, caves, and an ammunition cache. The early-morning strikes killed 20 militants, most of whom were reportedly Uzbeks. A Pakistani soldier was killed during the ground offensive that morning when an IED exploded (Dawn, 2014).

6 July 2014
Hundreds of Taliban fighters reportedly cut their hair and beards to flee the operation. According to IDPs, the Taliban disguised themselves in the weeks before the Pakistan Army operation. Although the militants advocated *sharia* in Pakistan and were contemptuous of Western culture, refugees said that in North Waziristan the militants enjoyed imported products.

7 July 2014
Pakistani Chief of Army Staff General Raheel Sharif visited North Waziristan. Sharif was received by Lieutenant General Khalid Rabbani, the Peshawar corps commander, and briefed by the general officer commanding for the operation. He commended the troops for their

determination, commitment and resolve, praising the progress achieved since the beginning of the operation.

8 July 2014

Pakistani Defence Minister Khawaja Asif said in an interview with Radio Pakistan that 400 militants and 20 Pakistani soldiers had been killed so far in the operation. Although he declined to provide a time frame for the operation's end, he expressed a resolve to finish as early as possible.[1] Airstrikes by the PAF killed 13 militants (including foreigners) in the Digaan area of North Waziristan, according to an ISPR press release. Seven militant hideouts were also destroyed.[1]

9 July 2014

Pakistani airstrikes targeted three hideouts, killing 11 militants in Shawal. Ground operations continued in Miramshah, which was 80-percent cleared.

10 July 2014

The Pakistani army invited local and international media to Miramshah to observe the militant facilities. Flags, weapons and explosive materials were shown. ISPR Director Major General Asim Bajwa and Operation Commander Major General Zafar Khan briefed the media on the operation's progress. They said that as many as 400 terrorists were killed and another 130 injured to date. Eleven IED factories were uncovered, and over 2,000 IEDs confiscated. Bajwa said that underground tunnels built by the militants (one a kilometer long) had also been discovered: "This (Miranshah) was one of their major base, we have dislodged them from here and now they are on the run, We have also discovered their command and control centre, which would definitely affect their capabilities" (the Express Tribune, 2014).

11 July 2014

Injured Taliban commander Adnan Rashid, al-Qaeda commander Mufti Zubair Marwat and Marwat's two guards were captured by Pakistani security forces in the Shakai Valley of South Waziristan when they tried to escape from cordoned-off North Waziristan, and were moved to an undisclosed location by army helicopter. Marwat was reportedly the brother of Mufti Sajjad Marwat, an al-Qaeda spokesman for Afghanistan and Pakistan. Rashid was planning to flee to Afghanistan. The arrest was confirmed by the TTP.

12 July 2014

Thirteen militants, primarily foreign, were killed in an early-morning PAF airstrike after firing rockets at a security checkpoint in Mir Ali; seven hideouts and an ammunition cache were also destroyed. The connection of the cleared area in Miramshah and the Miramshah-Dattakhei road continued. In Khar Warsak and Zartangi, security forces discovered six motorcycle IEDs, two vehicle IEDs, two 12.7 mm guns, one 14.5 mm gun, three vehicles and eleven explosive belts during the previous 24 hours. Two explosive-laden vehicles were also destroyed in airstrikes at Degan. Three militants, including one Uzbek, were arrested in Boya. Two suicide bombers were identified and chased, but they blew themselves up when encircled by security forces near the town. Eighteen militants were killed in PAF airstrikes in the Mosaki area, 25 kilometres (16 mi) east of Miramshah, and artillery shelling in the Kharkamar area, 30 kilometres (19 mi) west of Miramshah. "Six terrorist hideouts and a huge ammunition cache were destroyed and at least 13 militants, mostly Uzbek, were killed in the Saturday morning strikes, five militants were killed and two militant hideouts were destroyed by artillery fire in the Kharkamar area on Saturday night", a security official said.

13 July 2014

According to the ISPR, five hideouts in Mir Ali were targeted by the military and several militants were killed.

14 July 2014

After Miramshah was under control, a ground offensive was launched in Mirali (second-largest town in North Waziristan) and the nearby Boya area. Pakistani troops killed six militants (two of them suicide bombers), including Taliban commander Matiullah.

15 July 2014

In a media briefing, ISPR director general Major General Asim Bajwa said that 451 militants were dead and 88 hideouts destroyed. According to Bajwa, 26 soldierswere killed in the operation to date. According to ISPR, five Pakistani soldiers were killed in exchanges of fire in and around Mirali (including an officer, Captain Akash Rabbani). Two soldiers were injured, and 11 insurgents were killed.

16 July 2014

Thirty-five militants were killed in airstrikes in the Shawal area. According to an ISPR press release, "Today, early morning at least 35 fleeing terrorists were killed through aerial strikes in Shawal valley". The

Mir Ali ground offensive following the clearance of Miranshah continued, with airstrikes expected.

18 July 2014

House-to-house searches were conducted in areas of Mir Ali. Four militants were killed in an exchange of fire and 12 IEDs, an IED factory and caches of ammunition and foreign currency were seized.

19 July 2014

"The command and control system of terrorists was destroyed in North Waziristan," said Pakistani Defence Minister Khawaja Asif. "There is no place for terrorism in a democratic country."[1] The villages of Boya and Degan were cleared by Pakistani military, and the ground offensive continued in areas of Mirali. Corps Commander Lieutenant General Khalid Rabbani visited Mirali, Boya and Degan, meeting with the troops, and relief for IDPs was underway.

20 July

Twenty-eight militants were killed in airstrikes targeting six hideouts in the Shawal area of North Waziristan.

23 July 2014

Twenty militants, including foreigners, were killed by Pakistani airstrikes which destroying four hideouts in the Shawal tehsil of North Waziristan. The ground operation in Mir Ali continued, with an ammunition factory and foreign currency seized from Mir Ali Bazaar. According to ISPR, a disposal operation of mines and explosive materials was underway in Miramshah with six IED factories cleared by army engineers to date.

24 July 2014

Two Pakistani soldiers were killed in an evening IED explosion near Ghulam Khan, in North Waziristan (near the Pakistani-Afghan border).

26 July 2014

During the Mir Ali ground offensive, eight militants were killed and five hideouts were destroyed.[113]

27 July 2014

Pakistani security forces cleared 70 percent of Mir Ali and adjacent areas.

29 July 2014

The Pakistani Chief of Army Staff, General Raheel Sharif celebrated the Eid al-Fitr with army soldiers and IDPs in Baka Khel and Bannu.

30 July 2014
According to the official sources, a Pakistan Army check post was attacked by Afghan militants in Lower Dir. The cross-border attack involved 70-80 militants. In retaliation, at least seven militants were killed and nine others were injured (Dawn 2014).

August

2 August 2014
During the ground operation in Mirali, three militats were killed in fire-exchange. An ammunition dump was also seized.

4 August 2014
During ground offensive in Datta Khel area, Seven Uzbek militants were killed in fire-exchange, two soldiers identified as Subedar Mashkoor and Lans Naik Zaheer were also killed. Data Khel was cleared and the ground operation continued in Mirali and other areas.

5 August 2014
Airstrikes on 6 militant hideouts were conducted by the Pakistani military, killing at least 30 militants. According to ISPR, the raids were carried out in the Datta khel, Marshikhel and Kamsham areas. Mirali was also cleared and the ground operation in Mirali came to an end.

9 August 2014
The Pakistani Prime Minister Nawaz Sharif during National Security Conference in Islamabad said: "We pay tribute to the sacrifices of our armed forces in the war against terrorism and express solidarity with their families." Sharif added that the "intensity of the blowback of the military operation would be low."

14 August 2014
A Pakistani security forces camp in Miramshah was targeted with rockets by unknown militants. No loss of life occurred.

19 August 2014
48 militants were killed in airstrikes and shelling by gunship helicopters of the Pakistani military, destroying seven militant hideouts and several vehicles in North Waziristan and Khyber Agency.

30 August 2014
32 militants were killed and three hideouts destroyed by army gunship helicopters in the remote areas of North Waziristan. According to ISPR, 23 explosive laden vehicles and four ammunition dumps were also destroyed (Huffington Post, 2014).

September

3 September 2014

According to the ISPR, 910 militants had been killed so far in the operation. The ISPR statement added that eighty-two soldiers had also been killed (42 were killed in North Waziristan Agency), while 269 others have been injured. The Pakistan Armed Forces had cleared Miramshah, Mirali, Datta Khel, Degan, Boya areas of North Waziristan which were considered strongholds of terrorists.

8 September 2014

Ten militants were killed and five vehicles were destroyed by Pakistani gunship helicopters in Boya Degan, according to ISPR.

9 September 2014

A Pakistani soldier and six militants were killed in a counterattack during a clearing operation near Datta Khel. A civilian logistic staff member was also killed while providing supplies to the Pakistani forces.

10 September 2014

35 militants were killed when Pakistan bombed three militant hideouts in Datta Khel. Another 30 militants were killed when Pakistani fighter planes bombed two militant hideouts in Shawal.

12 September 2014

DG ISPR Asim Bajwa stated that the Pakistani Army was ready to go into remote areas to take down militants, if required. He added that the 10 militants behind the attack on Malala Yousafzai that took place on 12 October 2012 had been arrested, and those behind attack on Ziarat residency had also been arrested. He revealed that over 1,000 militants had been killed in North Waziristan including 45 hardcore militants, while 134 hardcore militants had also been arrested.

14 September 2014

Three Frontier Corps personnel were killed when a group of militants launched a rocket attack on a mountain fort in Spinwam area of North Waziristan near Afghan border.

15 September 2014

Fifteen militants were killed in fresh air strikes in North Waziristan. "Army Aviation Combat helicopters in precise strikes in Tabai area of North Waziristan Agency destroyed 10 explosive laden vehicles and 5 terrorists hideouts, 15 terrorists were killed," an ISPR statement said.

16 September 2014
Twenty militants were killed in Pakistani airstrikes in Khyber Agency's Tor Darra area targeting three militant hideouts and destroying two ammunition dumps. Another 11 militants were killed as a militant attack from across the border targeted Pakistani security forces in Dandi Kuch in the Spinwam area of North Waziristan. Pakistani troops also arrested one terrorist. Three Pakistani paramilitary soldiers were also killed in the gun-battle.

17 September 2014
At least 11 militants and three soldiers were killed on Tuesday when Afghan terrorists attacked a Pakistani border post in North Waziristan Agency from Afghanistan. Another 40 militants were killed in air raids on militant hideouts in North Waziristan. "Precision air strikes destroyed five terrorists' hideouts as well as ammunition dumps in village Nawe Kili and Zaram Asar in an area north of Dattakhel", ISPR said. "Many foreigners were among those killed in the air strikes", it added.

18 September 2014
23 militants were killed in airstrikes on militant hideouts. "Today in precise aerial strikes carried out on terrorists hideouts in Zerom, Ismail Khel, and Datta Khel in North Waziristan, 23 terrorists were killed," said ISPR.

20 September 2014
TTP spokesperson Shahidullah Shahid confirmed that a senior Commander Gul Hasan Afghani was killed in clashes with the Pakistan Army in the Boya area of the agency two days ago.[134] At least three militants were killed in an exchange of fire with the security forces in Boya area of the North Waziristan, the ISPR said. A Pakistan Army's naib subedar Muzzamil was also killed during the fire exchange.

23 September 2014
Twenty-three militants were killed in airstrikes in the Bangidar area of Ghulam Khanx in North Waziristan Agency, according to an ISPR press release. According to ISPR reports, nineteen militants were killed in air strikes on militant hideouts in Dandi Kachkol near Ghulam Khan, Gurbaz and Mana near Pasht Ziarat in North Waziristan.

28 September 2014
At least 15 militants were killed when Pakistan air force jets struck militant hideouts in North Waziristan. Foreign militants were also among the dead. Five militant hideouts were also destroyed (Shahzeb, 2014).

October

3 October 2014

Fifteen militants were killed by airstrikes in the Jamrud and Bara areas of Khyber Agency, and three militant hideouts were destroyed, ISPR said in a press-release. Over 1,200 militants had been killed in the operation so far.

5 October 2014

Despite the upcoming winter, the Pakistani military decided to continue the operation, dispelling an impression that the harsh weather may force military authorities to halt Operation Zarb-e-Azb, which has been ongoing since mid-June 2014. "Conducting operation in winters is certainly a challenge for the security forces but it is also a challenge for militants," a military official told the media.

12 October 2014

At least 11 militants were killed in military airstrikes targeting two militant hideouts in North Waziristan. In separate airstrikes in the Khyber Agency's Kuki Khel area, at least 10 militants were killed. Three militant hideouts were also destroyed, said an ISPR statement.

15 October 2014

Five people were killed and seven were injured in a suicide attack targeting members of a peace committee, in the Peer Mela area of Tirah valley.

16 October 2014

In an apparent response to the previous day's suicide bombing, 21 suspected terrorists were killed in airstrikes in the Khyber Agency's Tirah Valley, the Pakistani military said. Five terrorist hideouts were also destroyed.

17 October 2014

Eight militants were killed in the Aka Khel area of Khyber Agency, in clashes with Pakistani security forces. A militant hideout was also destroyed during the clashes.

21 October 2014

28 militants, including foreign militants, were killed in Pakistani airstrikes in the Datta Khel area of North Waziristan.

22 October 2014

DG ISPR Asim Bajwa, addressing the media after a friendly cricket match was played between the IDPs, accompanied by star cricketer Shahid Afridi and the Pakistani Army, said that Operation Zarb-e-Azb is going well, and that 1,100 militants have been killed so far. He refused to give a specific timeline of the operation's completion, and said that the army is working towards making a "comprehensive plan" to rehabilitate them.

25 October 2014

18 militants were killed in Pakistani airstrikes in Khyber Agency's Khyber district near the Afghan border.

27 October 2014

Pakistani officials claimed to have found evidence of "new sanctuaries" set up by the TTP and its affiliates in the Afghan territory, near the border with North Waziristan Agency.[148] In two separate incidents of aerial strikes, 33 militants were killed and nine militant hideouts were destroyed, according to the ISPR. "In early morning aerial strikes ahead of Dattakhel, 18 militants were killed. Later, in Gharlamai area another 15 militants were killed by gunship helicopters," the ISPR statement added.

29 October 2014

Twenty militants were killed, and eight others were injured in airstrikes conducted by the Pakistan Air Force in the Akka Khel area of the Khyber Agency (Omer, 2014).

November

6 November 2014

"Operation Zarb-e-Azb has helped disrupt Haqqani network's ability to launch attacks on Afghan territory", a senior commander for US and NATO forces in Afghanistan, Lieutenant General Joseph Anderson said in a Pentagon-hosted video briefing from Afghanistan. He added that the Haqqani network was now fractured too.

14 November 2014

Thirty militants including foreigners were killed in air strikes on militant hideouts in Datta Khel.

16 November 2014

27 militants including foreigners and commanders were killed in air strikes in Datta Khel militant hideouts. Last night, seven suspected terrorists were also killed during a search operation in Datta Khel. In the exchange of fire, three Pakistani soldiers were also killed while four were injured. "The Haqqani network and East Turkistan Movement have been eliminated from the area", commander of Operation Zarb-e-Azb Major General Zafarullah Khan said. He added that 1,198 terrorists were killed and another 356 were injured so far.

25 November 2014
Twenty terrorists, including members of the Haqqani network were killed in aerial strikes in North Waziristan(ISPR, 2014).

December

2 December 2014
32 militants, including foreigners, were killed in airstrikes in Datta Khel and Orakzai regions. In another incident, Pakistani military repulsed a pre-dawn militant attack on a military check post in the Sherin Dara area of Orakzai Agency.

3 December 2014
Fifteen militants were killed in aerial strikes targeting militant hideouts in North Waziristan(NWA, 2014).

6 December 2014
Pakistani special forces killed al Qaeda commander Adnan Gulshair el Shukrijumah. Several other militants were also killed during a targeted operation in the Wana subdivision of South Waziristan Agency. During the fore exchange, a Pakistani soldier Havildar Masood was also killed while another was critically injured.

8 December 2014
Airstrikes killed 30 militants, including important commanders in the Datta Khel area of North Waziristan. The killed were local commanders of Hafiz Gul Bahadur and his ally Sadiq Noor (Dunya News, 2014).

9 December 2014
It was rumoured that Hafiz Gul Bahadur, a militant commander, was killed in air strikes in Datta Khel district of North Waziristan Agency. A compound where 30 to 40 top commanders of Gul Bahadur's Shura Mujahideen were having a consultative meeting, was targeted killing all of them.

26 December 2014
Aerial strikes in Datta Khel killed 23 terrorists. An underground ammunition dump and tunnel system was also destroyed ([ISPR, 2014]).

31 December 2014
Pakistani air-strikes in the Shawal area of North Waziristan killed 23 terrorists (NWA, 2014).

(b). 2015

January

7 January 2015

Twelve militants were killed as gunship helicopters targeted militant hideouts in Datta Khel. In the air-strikes, four militant hideouts and seven vehicles were also destroyed.

16 January 2015

Director General Inter Services Public Relations (ISPR) Major General Asim Saleem Bajwa said security forces have killed 2,000 militants in North Waziristan so far. Bajwa added that 200 soldiers had been killed during the Operation Zarb-e-Azb and 800 others were injured.

25 January 2015

35 militants, including foreigners, were killed in air strikes targeting militant hideouts in Datta Khel.

27 January 2015

76 militants including foreigners were killed in the Pakistani air-strikes targeting militant hideouts in Datta Khel area of North Waziristan. 53 militants were killed in the first round and 23 militants were killed in the second round of the air-strikes. Six militant hideouts and an ammunition dump was destroyed.

February

20 February 2015

A Pakistan Army Major Zahid Shaheed was killed in Datta Khel with a sniper rifle. In Pakistani retaliation, 5 terrorists were killed.

April

22 April 2015

At least 35 militants were killed in aerial strikes in South Waziristan and Datta Khel area of North Waziristan, according to ISPR.

May

22 May 2015

Four Pakistani soldiers were killed and two others were injured in Dattakhel area of North Waziristan Agency at night after their vehicle was targeted by an improvised explosive device. In Pakistani counter-retaliation, the military launched a search-operation killing seven militants in the pursuing fire-exchange (The News Tribe, 2015).

June

28 June 2015

23 militants were killed in the Pakistani military air-strikes in the North Waziristan and Khyber agencies. Foreign militants were also among those killed in strikes which also targeted ammunition dumps.

July

5 July 2015

Seven Pakistani soldiers were killed in the Taliban attaks in North and South Waziristan. in Pir Ghr, Taliban attacked a Pakistani military convoy and killed two soldiers and wounded another three. Another two remote-controlled bomb attacks targeting military vehicles in North Waziristan and South Waziristan killed five soldiers and injured six. According to the Pakistani intelligence officials, the Pakistani military was facing fierce resistance in the final push towards Shawal valley.[170] In the pursuing fire-exchange and military's retaliation, 12 militants were killed in Datta Khel in the North Waziristan agency (The Express Tribune, 2015).

7 July 2015

A Pakistani soldier was killed in Taliban attack on an army bunker in the South Waziristan agency (Dawn, 2015).

American drone strikes

Drone strikes, which were halted for six months at the request of the Pakistani government, resumed for the operation. The following drone strikes took place during the operation.

In 2014

11 June 2014 – Two strikes in Miramshah killed 16 suspected militants and injured several others. These were the first drone strikes of 2014; the previous strike occurred on 25 December 2013 in the Qutab Khel area of Miramshah, killing four suspected militants.

18 June 2014 – At least six militants were killed in Miramshah.

10 July 2014 – A strike in the Datta Khel area killed seven militants and injured three others.

16 July 2014 - Four missiles were fired in a strike in the tehsil of Datta Khel, two on a house and two on a vehicle, killing twenty militants and injuring five.

19 July 2014 - Eleven militants, including two commanders, were killed in the tehsil of Madakhel, Data Khel, North Waziristan. Most of the militants belonged to the Punjabi faction of the Taliban.

6 August 2014 - A strike in Datta Khel killed six militants and injured two others.
24 Sept 2014: At least 8 people including Uzbek Militants were reportedly killed in a US drone strike Dattakhel tehsil of North Waziristan.
5 October 2014: At least 5 suspected militants were killed in a US drone strike in Shawal area of South Waziristan tribal region.
6 October 2014: At least 8 suspected militant were killed and several other injured in a U.S drone strike in Shawal district of North Waziristan.
7 October 2014: At least 3 suspected militants were killed in a U.S drone strike in North Waziristan region.
30 October 2014: A US drone strike killed at least 4, injuring several others in Birmal Tehsil of South Waziristan.
11 November 2014: A US drone strike in Doa Toi area of Datakhel tehsil in North Waziristan Agency killed 4 suspected militants (Dawn, 2015).
21 November 2014:Reportedly Five suspected militants including two commander of 'Qaedat al-Jihad in the sub-continent', a newly established branch of Al Qaeda were killed in a US Drone strike in Datakhel region of North Waziristan Agency.
6 December 2014: A US drone strike killed a key Al Qaeda leader Umar Farooq along with four others in Datakhel region of North Waziristan Agency.
26 December 2014: Two separate US Drone strikes in the Kund and Mangroti area of Shawal in North Waziristan Agency killed at least 7 suspected militants(Reuter, 2015).
In 2015
4 January 2015: A reportedly high-value unidentified Uzbek commander of Taliban's Gul Bahadur group was killed along with 8 others by a US drone strike in Shawal area of North Waziristan Agency.[186]
15 January 2015: A US drone strike reportedly killed 7 suspected militants in Wacha Dara area of Liddah Tehsil of South Waziristan Agency.
19 January 2015: A US drone strike killed 6 while injuring 4 others in the ShahiKhel area of North Waziristan's Shawwal tehsil.
28 January 2015: A US drone strike killed 7 while injuring another militant in the Shawal area of North Waziristan.
18 March 2015: A US drone strike killed a TTP commander Khawrey Mehsud along with 3 others in Shabak area of Kurram Agency.
12 April 2015: A U.S drone strike killed 4 suspected militants in North Waziristan.

16 May 2015: A drone strike Killed 7 to 13 militants in the Mana area of North Waziristan Agency.
18 May 2015: A U.S. drone strike killed 6 suspected militants in Zoye Narye Area of North Waziristan.
2 June 2015: Four suspected militants were killed in a drone strike targeting a vehicle in the Shawal area of North Waziristan.
6 June 2015: At least nine suspected militants were killed in a strike in Shawal's Zoya Saidgai area, considered to be a hideout of the Afghan Taliban (Press Release, 2015).

Yearly Development

On 13 June 2015, the Pakistani military reported progress in the operation in the course of a year. DG ISPR Asim Saleem Bajwa reported that 2,763 militants had been killed so far, including 218 terrorist commanders in 9,000 intelligence based operations (IBOs). "Some 837 hideouts of terrorists have been destroyed and 253 tonnes of explosives recovered so far during the operation", he said. The Army also recovered 18,087 weapons, including heavy machine guns, light machine guns, sniper rifles, rocket launchers and AK-47s . Bajwa said that thousands of terrorists were also arrested, their strongholds cleared and their communication infrastructure destroyed. "347 officers and soldiers of Pakistan Armed Forces have embraced martyrdom", he added. The year 2015 was declared to be "a year of victory" and the operation itself a "manifestation of the resolve to root out terrorism in the country" by the Pakistani Defence Minister Khawaja Asif (Dawn, 2015).

Final phase

In June 2015, it was reported that the Pakistani military was making preliminary preparations for the final phase of the operation. "Preliminary preparations for the final push have started, which will begin next month," a military official told media. The final phase will be focused in the areas south of Dattakhel towards the Afghan border through the mountainous and forested Shawal Valley, where a large number of militants are believed to have taken refuge after fleeing the operation in other parts of North Waziristan. "Operation Zarb-e-Azab moves to last few pockets close to Pak-Afghan border," military spokesman DG ISPR Asim Bajwa said. In July, the Pakistan Army Chief, General Raheel Sharif visited the military's forward-most positions near the Afghan border in North Waziristan He was briefed about the progress and future plans about the Operation Zarb-e-Azb and also met the Pakistani troops stationed there.

Sharif also visited South Waziristan and Wana agencies. It was reported that the Pakistani military had just completed the preliminary prepatrations for final phase. In the area around Shawal, peaks were cleared (Pakistan Today, 2015).

Management of displaced civilians

As a result of the operation, 929,859 displaced civilians (from 80,302 families) were registered by Pakistani authorities as of 14 July. Financial support, relief goods and food packages were being distributed and 59 donation points were established across Pakistan by the army. On 10 July, the Foreign Office of Pakistan said that the rehabilitation of internally displaced persons was an internal matter and reiterated that Pakistan had not requested international assistance. "We have very clear instructions from the prime minister [to not seek external assistance], Pakistan has neither made nor intends to make a request for international assistance. It has been made very clear that all expenditure related to temporarily displaced Pakistanis will be met from our own resources", Pakistani Foreign Office spokesperson Tasnim Aslam said. However, it was reported that the United States allocated $31 million for IDPs and an additional $9.3 million for health, hygiene, water and sanitation for IDPs and livestock. It was also reported that the United Arab Emirates government allocated $20.5 million in IDP humanitarian aid. In February 2014, the Pakistani Finance Minister Ishaq Dar told a visiting US Senator Jacl Reed tjat the cost of the operation so far had reached 40 bilion (US$390million) and could go as high as 130 billion (US$ 1.3 billion) (Walker, 2015).

Return of the IDPs

On 8 December 2014, the military approved the phased repatriation of the IDPs. The Pakistani army chief said the early return of the IDPs was his top priority. According to senior military officials, civilian authorities were directed to plan the return of the IDPs to their homes in areas which have been cléared of terrorists. On 31 March 2015, the repatriation started. On the first day, 219 IDPs belonging to 62 families left Bannu for Spinwam and Shahmeri in North Waziristan. In the first phase till 24 April, about 1,200 families were to return to their homes in Sinwam, Shamiri, Mirali and Bubali areas of NWA. Each family was given 25000 (US$ 250) as cash assistance and 10000 (US$ 98) as transportation expenses at Mirzail. Food ration for six months and non-food items would also be given to each household. For effective disease control, children under five years of age were administered anti-polio vaccines and under 10 years of age were administered anti-measles vaccines (The Nations, 2015). As of 4 May 2015, only 230 displaced families had returned since the launch of the repatriation programme. "The civil administration can send entire displaced population back to their homes within a month if the area is de-notified as conflict zone," said an official dealing with IDPs repatriation. According to the official sources, the cut-off date for the return of IDPs was December 2016.

Reaction

(a). Domestic

The decision by the Pakistani military to launch a comprehensive operation was widely supported, with journalists, opinion-makers, politicians and other social-media users commending the operation.

Pakistan Tehrik-e-Insaf

PTI chairman Imran Khan endorsed the military operation in North Waziristan as it became clear that the Taliban were not seriously negotiating. A week before, reiterating his party's stance on peace talks with militants, Imran said that an offensive in North Waziristan would unite militant forces against the Pakistani state. "Conducting such a military operation when most of the groups in NWA want talks is suicidal," Imran said in a statement, adding that most groups in North Waziristan desired peace talks with the government. The PTI position changed as it became clear that negotiations were fruitless(Dawn, 2015).

Jamat-e-Islami

Jamat-e-Islami (JI), one of Pakistan's leading religious parties, continued to oppose any operation in North Waziristan. JI leader Siraj-ul-Haq urged the government to keep the option of negotiations with the Taliban.[215] He warned that a military operation in North Waziristan would trigger a massive human tragedy, saying that it was the duty of Prime Minister Nawaz Sharif to consider the views of the nation and its leadership (inside and outside Parliament) before making a crucial decision affecting national security (Long War, 2015).

Muttahida Qaumi Movement

According to Muttahida Qaumi Movement senator Babar Khan Ghauri, "This is a commendable decision by the government. We have been repeatedly telling the current government that instead of engaging the terrorists in dialogue, government should act against these elements. Karachi has a number of terrorists and it might suffer from a blow back;

this should be tackled so Karachi does not have to suffer on account of this." MQM head Altaf Hussain said, "I welcome this operation and I am glad that government is supporting the armed forces, those who have not backed the operation must realize that it is a matter of national security. I appeal to them to come on same page by setting aside their political compulsions" (DND, 2015).

Awami National Party

Awami National Party (ANP) member Zahid Khan said, "We also held a dialogue previously (during our government) but that did not produce effective results. We wanted peace and we were okay if that came through dialogue but unfortunately that could not happen. This time, knowing from our experience, we had cautioned the government that [the] dialogue approach would not work. Government should have taken the parliament into confidence before launching the operation but it didn't."

Local tribesmen

North Waziristan tribal elders assured their support for Operation Zarb-e-Azb, according to a statement released by ISPR director-general Major General Asim Bajwa. "Many tribal elders from around Miranshah, Mir Ali, Datta Khel assure support to army operation," Bajwa tweeted. "The tribesmen have assured the army that they would not let the militant to return to the area."

Sunni Ulema Board

On 22 June 2014, more than 100 Islamic scholars issued a joint fatwa in support of the operation, calling it a jihad: "Crushing of the attempts to disrupt peaceful atmosphere in a Muslim state is jihad" (The News, 2015).

(b). International

Afghanistan – Afghan Ambassador to Pakistan Janan Mosazai stated that his government would provide "every possible assistance" to defeat the militants in the operation.

United Arab Emirates – Interior minister Saif bin Zayed Al Nahyan said that his government would co-operate with Pakistan in the war against the extremists.

United Nations – In a statement, the United Nations Office for the Coordination of Humanitarian Affairs (OCHA) estimated that as of 23 June more than 450,000 people were internally displaced from the war-torn region. Other UN agencies, such as the UNHCR, agreed to provide tents and other facilities to the camps. The World Health

Organization (WHO) provided medicines and vaccines to the IDPs to avert a polio outbreak.

United States – The US supported military operations against Taliban militants, a spokesman from the US Embassy in Pakistan said on 16 June, and the US supported every Pakistani step taken for the establishment of peace. The United States had pressured Pakistan for a military operation in North Waziristan for years, and the US Congress linked military assistance to Pakistan for the next fiscal year with military operations in North Waziristan in June 2014. Rear Adm. John Kirby, the Pentagon press secretary, said that the Pentagon was unaware of Pakistan's decision to launch a new offensive in North Waziristan: "The Pakistan military and the government understand the threat, and they continue to go after that threat" (Obama, 2015). On 5 November 2014, Lt. Gen. Joseph Anderson, a senior commander for US and Nato forces in Afghanistan, said in a Pentagon-hosted video briefing from Afghanistan that the Haqqani network is now "fractured" like the Taliban. "They are fractured. They are fractured like the Taliban is. That's based pretty much on the Pakistan's operations in North Waziristan this entire summer-fall," he said, acknowledging the effectiveness of Pakistan's military offensive. "That has very much disrupted their efforts in Afghanistan and has caused them to be less effective in terms of their ability to pull off an attack in Kabul," Anderson added (Telegraph, 2015).

China – Chinese Foreign Minister Wang Yi said that terrorism was a problem common to China and Pakistan, since militants were the enemy of both countries, adding that China fully supported the operation (Business Standard, 2015).

Russia – Chairman of the State Duma Sergey Naryshkin commended the operation, while the Pakistani Army Chief, General Raheel Sharif was on an official visit to Russia. "We will stand by Pakistan in its fight against terrorism and extremism for stability in the region. Our relations are independent, more consistent and will further grow" (Russa, 2015).

References

1. BBC. (2014, 26 March).First day of Pakistan Taliban peace talks concluded.. Archived from the original on 30 June 2014. Retrieved 15 June 2014.
2. Business Standard. (2015, 14 June). Pak to launch 'final phase' of Operation Zarb-i-Azb next month. Retrieved 17 June 2015.
3. Dawn. (2014, 30 June 2014).Pakistani Taliban announce month-long ceasefire – Pakistan. Archived from the original on 30 June 2014. Retrieved 15 June 2014.
4. Dawn. (2014, 3 January). Hafiz Gul Bahadur group targeted in N Waziristan drone strike. Zahir Shah Sherazi. Retrieved 3 January 2015.
5. Dawn. (2014, 15 June).Zarb-e-Azb operation: 120 suspected militants killed in N Waziristan. Archived from the original on 13 July 2014. Retrieved 16 June 2014.
6. Dawn. (2015, 15 January). US drone strike kills seven in South Waziristan. Zahir Shah Shirazi. Retrieved 16 January 2015.
7. Dawn. (2015, 19 January). "US drone attack in North Waziristan kills six". Zahir Shah Shirazi. Retrieved 19 January 2015.
8. Dawn. (2015, 28 January). "US drone strike kills six in North Waziristan". Zahir Shah Shirazi. Retrieved 28 January 2015.
9. Dawn News. (2015,17 May) US drone attack in NWA kills six. Retrieved 14 June2015.
10. Dawn News.(2015,14 June).Conclusive phase of Zarb-i-Azb next month. .Archived from the original on 17 June 2015. Retrieved 17 June 2015.
11. Dawn News. (2015, 1 April).IDPs' return to North Waziristan begins. Archived from the original on 23 May 2015. Retrieved 23 May 2015.
12. Dawn News. (2015, 4 May). Return of North Waziristan IDPs continues at snail's pace. . Archived from the original on 23 May 2015. Retrieved 23 May 2015.
13. Deal with Russia for purchase of Mi-35 attack helicopters on the cards. Retrieved19 June 2015.
14. DND. (2015, 13 June). 2,763 terrorists, 347 Pakistan army personnel killed in Zarb-e-Azb so far: ISPR. Archived from the original on 14 June 2015. Retrieved 14 June2015.
15. Dunya News. (2014, 30 June). Negotiations end, Pak Army formally launches operation Zarb-e-Azb against terrorists; Dunyanews.tv. Archived from the original on. Retrieved15 June 2014.
16. Express Tribune. (2014, 2 July). Haqqani fighters won't be spared either. Archived from the original on 2 July 2014. Retrieved 2 July 2014.
17. Gul, Pazir (2014,16 June). Seven killed in clash between militant groups. Dawn. Archived from the original on 30 June 2014. Retrieved 17 June 2014.
18. Huffington Post. (2014,15 June). Associated Press.. Archived from the original on 30 June 2014. Retrieved 23 June 2014.
19. Indian Express. Suicide bombing? At least 55 killed, over 200 injured in Wagah border bomb blast..com.
20. IMU announces death of emir, names new leader. (2014, 4 August).The Long War Journal.
21. ISPR. (2014, 28 June). Archived from the original on 28 June 2014. Retrieved 28 June 2014.

22. Jets strafe militant hideouts; 16 killed". 30 June 2014. Archived from the original on 30 June 2014. Retrieved 30 June 2014.
23. Key TTP commander Adnan Rashid arrested in S Waziristan: sources. Dawn (Newspaper). 16 July 2014. Archived from the original on 16 July 2014. Retrieved16 July 2014.
24. Khan, Tahir. (2014, 20 June). Anti-terrorism cooperation: Islamabad asks Kabul to extradite Fazlullah. The Express Tribune.
25. Long War Journal. (2015, 6 June). US kills Afghan Taliban fighters in drone strike in Pakistan. . Archived from the original on 14 June 2015. Retrieved 14 June 2015.
26. Militant ambush in NWA kills two security personnel". 1 July 2014. Archived fromthe original on 2 July 2014. Retrieved 2 July 2014.
27. News Week Pakistan.(2014,8 July), Another 13 Militants Killed In North Waziristan, Says ISPR. Archived from the original on 8 July 2014. Retrieved 8 July 2014.
28. NWA operation: Seven more militants killed; toll rises to 187-Pakistan. Dawn (Pakistan). Archived from the original on 16 June 2014. Retrieved 16 June2014.
29. North Waziristan: 12 militants killed in air strikes. Dunyanews.tv. Archived from the original on 17 July 2014. Retrieved 16 June 2014.
30. Operation Zarb-e-Azb updates archive. ISPR. Retrieved 30 June 2014.
31. Omer Farooq Khan. "Pak military launches onslaught against Taliban". The Times of India. Archived from the original on 17 July 2014. Retrieved 16 June 2014.
32. Obama 2015 Pakistan drone strikes. thebureauinvestigates.com. Retrieved 16 April2015.
33. Press Releases of 16 June, ISPR, 16 June 2014, archived from the original on 20 June 2014, retrieved 20 June 2014.
34. Pakistan Tribe. (2014, 24 December). Retrieved 24 December 2014.
35. Press Releases of 24 June, ISPR, 24 June 2014, archived from the originalon 24 June 2014, retrieved 24 June 2014.
36. Press Releases of 25 June, ISPR, 25 June 2014, archived from the original on 28 June 2014, retrieved 25 June 2014.
37. Press Release. Pakistan Condemns US Drone Strike in North Waziristan.mofa.gov.pk. Ministry of Foreign Affairs Government of Pakistan. Retrieved 10 June2015.
38. Pakistan Today. (2015, 6 June). At least nine killed in NWA drone strike. Archived from the original on 14 June 2015. Retrieved 14 June 2015.
39. Reuters. (2015, 12 April).Drone strike in northwest Pakistan kills four suspected militants. Javed Hussain and Saud Mehsud. Retrieved 16 April 2015.
40. Russia stands with Pakistan in fight against extremism, Army chief told. Retrieved19 June 2015.
41. Saleem, Farukh. (2014, 14 October). India disappointed by Zarb-e-Azb's success The News International.
42. Shah, Saeed (2014, 15 June). "Pakistan Operation Targets Waziristan Militants". The Wall Street Journal. Retrieved 16 June 2014.
43. Shahzeb Jillani (2014, 6 February). Pakistan enters peace talks with Taliban. BBC. Archived from the original on 30 June 2014. Retrieved 15 June 2014.

44. The Express Tribune. (2015, 2 June). US drone strike kills four suspected militants in North Waziristan. . Archived from the original on 14 June 2015. Retrieved 14 June2015.
45. The Express Tribune. (2015, 16 June),A hard-fought year. Archived from the original on 17 June 2015. Retrieved 17 June 2015.
46. The Nation. (2015, 3 July). COAS visits forward most locations near Pak-Afghan border. Archived from the original on 3 July 2015. Retrieved 3 July 2015
47. Times of India. (2014, 23 November).
48. The News International. (2014, 15 June). Pakis: n.
49. The Express Tribune. (2014, 14 July) Air raids flatten 5 militant hideouts. . Archived from the original on 14 July 2014. Retrieved 14 July 2014.
50. The News. (2014, 2 July).
51. The Express Tribune. (2014, 10 June). Operation Zarb-e-Azb updates. Desk, Web. Archived from the original on 13 July 2014. Retrieved 15 June 2014.
52. The News Tribe. (2015, 5 July). 12 terrorists killed by security forces in North Waziristan". Archived from the original on 5 July 2015. Retrieved 5 July 2015.
53. The Express Tribune. (2015,7 July). Taliban kill Pakistani soldier in attack on army bunker. Archived from the original on 7 July 2015. Retrieved 7 July 2015.
54. The Express Tribune. (2015,19 February). Seeking assistance: Zarb-e-Azb to cost $1.3b, Dar tells US legislator. Archived from the original on 24 May 2015. Retrieved24 May 2015.
55. The Express Tribune. (2014, 8 December). Winding up the offensive: Army gives nod for return of IDPs to N Waziristan. . Archived from the original on 23 May 2015. Retrieved 23 May 2015.
56. The News. (2015, 18 June)."Russia praises Operation Zarb-e-Azb". . Archived from the original on 18 June 2015. Retrieved 18 June 2015.
57. Telegraph. (2014, 26 December). Yusufzai, Ashfaq "Mastermind of Peshawar school attack killed, Pakistan Bureau, 2014. Telegraph, Pakistan Bureau. Retrieved20 June 2015. Pakistan says it has killed the alleged mastermind of a school attack in Peshawar that claimed the lives of 132 children.
58. Walsh, Declan (2014,16 June). In Drive Against Militants, Pakistani Airstrikes Hit Strongholds. The New York Times. Archived from the original on 30 June 2014. Retrieved 17 June 2014.
59. Walker, Lauren. Report: Drone Strikes Kill Dozens in Afghanistan, Pakistan and Yemen in May. newsweek.com. NEWSWEEK. Retrieved 10 June 2015.
60. Zarb-e-Azb: 327 militants killed so far, tells DG ISPR". ARY News. 26 June 2014. Archived from the original on 27 June 2014. Retrieved 27 June 2014.
61. Zarb-e-Azb: 80 percent of Miranshah cleared; 11 more killed in jet bombing. Daily Times. 10 July 2014. Archived from the original on 10 July 2014. Retrieved10 July 2014.

National Action Plan

1. Execution of convicted terrorists
2. Establishment of special trial courts

   Ensure no armed militias are allowed to function in the country
3. Strengthening and activation of NACTA
4. Countering hate speech and extremist material
5. Choking financing for terrorists and terrorist organisations
6. Ensuring against re-emergence of proscribed organisations
7. Establishing and deploying a dedicated counter-terrorism force
8. Taking effective steps against religious persecution
9. Registration and regulation of madrassas
10. Ban on glorification of terrorism and terrorist organisations through print and electronic media
11. FATA Reforms
12. Dismantling communication networks of terrorist organisations
13. Measures against abuse of internet and social media for terrorism
14. Zero tolerance for militancy in Punjab
15. Taking the ongoing operation in Karachi to its logical conclusion
16. Balochistan reconciliation
17. Dealing firmly with sectarian terrorists
18. Policy to deal with the issue of Afghan refugees
19. Revamping and reforming the criminal justice system

Time-line of Operation Zarb-e-Azb

Date	Achievement/Development	Reference
June 15, 2014	DG ISPR has said that on the directions of the Government, Armed forces of Pakistan have launched a comprehensive operation against foreign and local terrorists who are hiding in sanctuaries in North Wazirastan Agency. The operation has been named Zarb-e-Azb.	No PR124/2014-ISPR
June 15, 2014	8 hideouts of Terrorists in NWA were destroyed killing 105 terrorists, most of them Uzbek foreigners.	No PR125/2014-ISPR
June 16, 2014	6 x soldiers embraced shahadat while 3 are injured due to IED explosion between Afghan border and Ghulam khan in North Wazirastan Agency today.	No PR128/2014-ISPR
June 16, 2014	7 Terrorists were killed in exchange of fire and 2 soldiers also embraced shahadat.	No PR129/2014-ISPR
June 16, 2014	Chief of Army Staff General (COAS) Raheel Sharif visited Corps Headquarters Peshawar today. COAS emphasized that all terrorists along with their sanctuaries must be eliminated without any discrimination.	No PR130/2014-ISPR
June 17, 2014	6 x Terrorist's hideouts including a training camp and an IED making factory around Hasokhel, North Waziristan Agency were destroyed by jet aircrafts 25 x foreign and local terrorists were killed in the strikes 3 x terrorists were killed while fleeing from the cordoned off area in Miranshah. 1 x soldier got injured in exchange of fire.	No PR131/2014-ISPR
June 19, 2014	15 x terrorists were killed in Zartatangi mountain heights, 8 x Uzbeks were killed around Miranshah	No PR135/2014-ISPR

June 20, 2014	3 x Terrorists hideouts in area Qutab Khel wrere destroyed and 12 x Terrorists including foreigners were killed in the strikes. 24 x Suspects who were trying to flee in the garb of IDPs have been apprehended at various strangulation check posts in Mirali and Miranshah.	No PR136/2014-ISPR
June 21, 2014	30 x terrorists were killed in targeted strikes by jet aircrafts in Khyber and North Waziristan Agencies , 3 x hideouts were destroyed in Hassu Khel in North Waziristan Agency killing 20 x Terrorists.	No PR138/2014-ISPR
June 23, 2014	8 x Terrorist's hideouts around Mirali, were destroyed. 10 x terrorists were killed and 2 x soldiers embraced shahadat	No PR142/2014-ISPR
June 24, 2014	27 x terrorists were killed in jet aircraft strikes in Mir Ali and 11 x Terrorist's hideouts were destroyed. 20 x Terrorists were killed in early morning air strikes on terrorist's hideouts in Khyber Agency. 12 x terrorist's hideouts were destroyed. 2 soldiers and a civilian embraced shahadat when nearby roof of a building collapsed due to explosion.	No PR143/2014-ISPR
June 25, 2014	5 x Terrorist's hideouts were destroyed by PAF jet aircrafts this afternoon around Mirali killing 13 x terrorists. 12 x Terrorists surrendered to security forces in North Waziristan Agency.	No PR145/2014-ISPR
June 28, 2014	19 terrorists surrenederd, PAF Jet aircrafts destroyed 6 confirmed terrorists' hideouts on the outskirts of Mirali killing 11 x Terrorists. Terrorist's concentrations were targeted early morning today by integrated fire of Artillery, Tanks and Heavy weapons outside Miranshah in North Waziristan Agency killing 7 x terrorists. 3 Terrorists were arrested by security forces.	No PR148/2014-ISPR

	TTP commander of Miranshah, Commander Umer has been killed by security forces last night on the outskirts of Miranshah.`	
June 30, 2014	15 x terrorists killed and 3 x soldiers got injured in exchange of fire. Since start of operation Zarbe Azb on 15 June, 376 x terrorists have been killed while 19 have surrendered to the security forces. 61 x terrorists Hideouts have been destroyed in the operation during last 15 days. 17 x Soldiers have embraced shahadat in the ongoing Operation Zarbe Azb	No PR150/2014-ISPR
July 5, 2014	5 terrorists hideouts, caves and huge cache of arms and ammunition were destroyed and score of terrorists killed in early morning strikes, silencing the firers. Most of the terrorists killed in strikes are Uzbeks.	No PR157/2014-ISPR
July 12, 2014	13 x terrorists and destroying 7 hideouts. A huge cache of arms and ammunition was also destroye+6d. Consolidation of cleared area in Miranshah and Miranshah- Dattakhel road continues. 6 prepared Motorcycle borne IEDs, and 2 Vehicle Borne IEDs, 2 Guns 12.7 mm, 1 Gun 14.5 mm, 3 Vehicles, 11 suicide jackets and huge cache of arms and ammunition were recovered during last 24 hours in area Khar Warsak and Zartangi.	No PR161/2014-ISPR
July 15, 2014	2 x soldiers including an officer Capt Akash Rabbani embraced Shahadat and 2 x soldiers were injured 3 x soldiers embraced Shahadat in Mirali during day and 7 x terrorists were killed. Today a total of 5 x soldiers have embraced Shahadat in Mirali and 11x Terrorists were killed.	No PR162/2014-ISPR
July 16, 2014	Early morning 35 x fleeing terrorists were killed through aerial strikes in shawal valley.	No PR163/2014-ISPR

July 19, 2014	4 x terrorists were killed in Mirali, 12 IEDs were neutralized last night alone, while an IED making factory, huge cache of arms and ammunition and foreign currency were also recovered. Meanwhile strangulation and targeting of terrorists and their hideouts in other areas in North Waziristan Agency is continuing.	No PR164/2014-ISPR
July 23, 2014	4 x isolated terrorist hideouts were destroyed, killing 20 x local and foreign terrorists.	No PR167/2014-ISPR
July 27, 2014	Operation ZarbeAzb continues to achieve its objectives and is progressing well. After clearance of Miranshah, Boya, Degan from terrorists, Mirali Town alongwith its neighboring villages, the second largest stronghold of terrorists has been 70% cleared by security forces till this morning. House to house search continues. 2 huge ammunition factories with ammunition stock were recovered yesterday.30 x barrels filled with explosives and chemicals used for making IEDs were also recovered from 2 x large underground tunnels during Clearance of Mirali.	No PR169/2014-ISPR
August 4, 2014	During clearance of surrounding areas of Mirali in Shahbaz Khel 75 Rockets, large number of IEDs and suicide jackets were recovered today. .	No PR172/2014-ISPR
September 3, 2014	Since start of the operation 910 x Terrorists have been killed. As many as 27 x IEDs Factories, 1x Rocket and 1 x Ammunition factory have been recovered and destroyed.	No PR187/2014-ISPR
September 12, 2014	The entire gang involved in the murder attempt of Malala Yousafzai, Kainat and Shazia has been busted. The gang known as "Shura" comprised 10 terrorists and all of them have been arrested. The group was part of TTP.	No PR197/2014-ISPR

November 16, 2014	27 Terrorists including some of their important commanders and foreigners were killed in Dattakhel, North Wazirsstan Agency today.	No PR252/2014-ISPR
November 17, 2014	10 x Terrorists surrendered to security forces along with their weapons and ammunition in Bara area of Khyber Agency. Since start of operation Khyber 1 total of 350 terrorists including more than 20 important commanders have surrendered to security forces.	No PR255/2014-ISPR
June 1, 2015	12 Terrorists were killed by security forces(SF)	No PR156/2015-ISPR
June 1, 2015	10 terrorists including foreigners were killed	No PR157/2015-ISPR
June 8, 2015	19 terrorists were killed including 5 of their commanders , 7 soldiers embraced shahadat.	No PR163/2015-ISPR
June 13, 2015	One Year Update of Zarb-e-Azb 1. Zarb-e-Azb: 1 year since start, significant achievements in FATA specially NWA&Khyber. Strong holds, Communication Infrastructure, sanctuaries largely cleared. 2. Zarb-e-Azb: Op moves to last few pocketts close to Pak-Afgn border. So far 2763 terrorists killed, 837 hide out destroyed. 253 tons explosive recovered. 3. Zarb-e-Azb: 9000 IBOs apprehended thousand terrorists, abettors & killed 218 hard core terrorists in cities. 347 offrs, soldiers embraced shahdat. 4. Zarb-e-Azb: 18087 weapons including Heavy Machine Guns, Light Machine gun, sniper rifles, rocket launcher, AK-47s searched out of terrorists hide outs from NWA alone. Base denied	No PR171/2015-ISPR

	5. Zarb-e-Azb: Whole nation came together to totally eliminate terrorism from Pak Media, civ society playing +ve role to help end menace. 6. Zarb-e-Azb: Will continue ops for terror free Pak.Few effects; improved envmt; Chinese Pres & other dignitaries' visits, Pak-Zimb cricket, CPEC.	
June 28, 2015	23 terrorists including foreigners were killed	No PR178/2015-ISPR
July 5, 2015	12 terrorists were killed and 4 soldiers also embraced shahadat in exchange of fire.	No PR182/2015-ISPR